Deep Tissue Insights: Unveiling Chemical Composition with Coherent Raman Scattering Microscopy

Anton

Copyright © 2024 by Anton

All rights reserved. No part of this book may be reproduced in any man-ner whatsoever without written permission except in the case of brief quotations embodied in critical articles and reviews.
First Printing, 2024

Table of Contents

Chapter 1

Introduction to Nonlinear optical microscopy

1. Introduction

In the realms of biological and geological research, nonlinear optical microscopy (NLOM) techniques have proven to be effective imaging tools [1-22]. The contrast mechanism in NLOM is based on nonlinear interactions between light and sample. Focusing a short laser pulse (ps-fs) by an objective into some samples induces a nonlinear polarization effect. The physics behind the nonlinear optical process can be understood in term of the Taylor expansion of the polarization as the function of the input electric fields. For example, when the polarization depends on the second power of the input electric field, then under some conditions (phase matching) the second harmonic of the input electric field will be generated. The generated nonlinear signal could be used as a contrast mechanism in imaging. The contrast mechanism in traditional microscopy (linear) is primarily based on linear interactions between light and sample, such as absorption, scattering, and refraction. Adopting NLOM can provide a number of advantages, including 3D imaging without the necessity of a confocal pinhole. This could increase the signal-to-noise ratio and make it easier to build miniaturized microscopes for in-vivo applications. NLOM commonly uses light with a near infrared wavelength. In most tissues, this light is only weakly absorbed, and nonlinear absorption occurs only in the focal volume. Since the first demonstration of two-photon excited fluorescence microscopy (TPEF) in 1990, NLOM has grown at a remarkable pace [9]. Second harmonic generation (SHG) [10], and third harmonic generation (THG) [11] techniques have also been proved to be effective. Although TPEF, SHG, and THG can provide contrast without the need for sample labelling, thery are not, in general, a chemically specific imaging method. Infrared and confocal Raman (spontaneous) microscopy can be used to image with chemical specificity, although each have limits [12-13]. Infrared microscopy has low spatial resolution due to the long wavelength. The difficulty of constructing optics for long wavelengths exacerbates the issue. Despite the fact that Raman microscopy does not have the same difficulties as infrared microscopy, because molecular Raman cross sections are very small, signal collection requires long integration times, which reduces scan speed. Furthermore, depth profiling and fluorescence background suppression in Raman microscopy can be difficult to achieve [14]. Light damage is a nonlinear process in some materials [15], which is a disadvantage of NLO microscopy. It can be challenging to choose the right laser source because of this. Another disadvantage is that the laser source used is significantly more complex and costly than those required for linear optical microscopy.

Coherent Raman Scattering (CRS) microscopy is a type of non-invasive, label-free, and non-destructive imaging modality based on the third-order non-linear susceptibility [1-11]. The Raman scattering effect is used in CRS microscopy. It is possible to use the vibrational modes of molecules in relation to chemical bonds as a contrast mechanism for identification, analogous to fingerprints. In CRS, two input laser pulses are required, one at frequency ω_p (pump) and one at frequency ω_s

(Stokes). As it is a stimulated process, the molecules in the focal volume oscillate in phase as a vibrational coherence which enhances the radiated signal by many orders of magnitude over the spontaneous Raman effect, particuarly when the frequency difference between the input fields equals the resonant frequency of a vibrational mode of the molecules in the sample. Coherent Anti-Stokes Raman Scattering (CARS) and Stimulated Raman Scattering (SRS) are the two most prevalent CRS implementation approaches. The energy diagrams for the SHG, THG, TPEF and SRS processes are shown in Fig. 1.1.

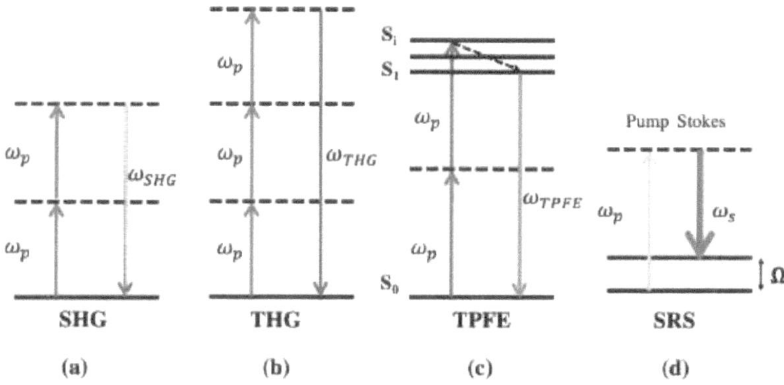

Fig. 1.1. Energy level diagrams of four different nonlinear optical processes. Dashed lines are virtual levels and solid lines are real levels. (a) Second harmonic generation, (b) Third harmonic generation, (c) Two photon fluorescence emission and, (d) Stimulated Raman Scattering. Image contrast could be achieved using nonlinear optical techniques.

In Fig. 1.1, the SHG, THG and TPFE processes are parametric. In a parametric process, the initial and the final quantum states of the system are the same. Consequently, there is no energy or momentum exchange between input fields and the quantum states of the system. On the other hand, in a nonparametric process like SRS, the initial and the final quantum states of the system are different. Both parametric and nonparametric processes can be described using the optical susceptibilities. We employed the SRS process as an image contrast mechanism in this thesis. The classical theory of Raman scattering will be addressed first in the following sections, followed by a discussion of stimulated Raman scattering.

1.1 Classical theory of Raman Scattering

The interaction of light and matter can be explained by absorption and scattering at thermodynamic equilibrium [40]. Molecular scattering is categorized in two ways: (i) elastic (Rayleigh) and (ii) inelastic (Raman). The difference is that with elastic scattering, the scattered light frequency is the same as the incoming light frequency. In Raman, however, the frequency of scattered light differs from the frequency of incident light. The Raman effect which was first reported in 1928 by Raman and Krishnan [16] and, independently, by Landsherg and Mandelstam. They spectrally filtered sunlight and then focused it into a sample [40].

1.1.1   Induced polarization

How light interacts with molecules can be understood via the interaction of electromagnetic waves with charged particles within a substance or molecule. The electric field associated with the visible and near-infrared ranges of the electromagnetic spectrum oscillates at 100s of THz: the heavy atomic nuclei cannot adiabatically follow the driving frequency [17, 18]. On the other hand, the much lighter electrons are often able to follow the rapid oscillations of the driving electromagnetic field. Consequently, the bound electrons are slightly displaced from their field-free positions, resulting in an induced oscillating electric dipole moment as suggested by Eq. 1.1:

$$\mu(t) = -e.r(t). \tag{1.1}$$

Where, e is charge of an electron and r represents displacement from the equilibrium (field-free) position. The microscopic polarization is obtained by summing over the N electric dipoles per unit volume, shown below:

$$P(t) = N\mu(t). \tag{1.2}$$

In the limit of a weak applied electric field, the microscopic polarization is linearly proportional to the electric field. This can be seen in Eq 1.3:

$$P(t) = \epsilon_0\chi E(t). \tag{1.3}$$

ϵ_0 is electric permittivity in vacuum and χ is linear susceptibility of the material. Almost all linear optical phenomena are explained by the linear relation of polarization with electric field. When the electric field becomes greater, electrons are pushed further away from equilibrium, and the binding potential between electrons and nuclei shifts from harmonic to anharmonic [17]. As a result, the relationship between induced polarization and electric field is no longer strictly linear. If the anharmonic contributions to the harmonic potential are relatively small, the displacement $r(t)$ can be expanded with a Taylor series. The polarization can therefore be represented as a power series in the electric field, as shown in Eq 1.4.

$$P(t) = \epsilon_0[\chi^{(1)}E + \chi^{(2)}E^2(t) + \chi^{(3)}E^3(t) + \cdots] \tag{1.4}$$

The first term in Eq 1.4 contains all linear optics, which we shall utilize to describe spontaneous Raman scattering. The second term describes the image contrast mechanism in second harmonic microscopy, and the third term will be used to explain the coherent Raman effect, which includes CARS and SRS.

1.1.2   Classical description of spontaneous Raman scattering

As explained in 1.1.1, the induced polarization is directly proportional to the electric field in week field regime. The polarizability $\alpha(t)$, connects the induced polarization to the electric field [17]. The induced dipole moment, as demonstrated in Eq 1.5, can also be utilized to explain the relationship between nuclear motion and electric field.

$$\mu(t) = \alpha(t)E(t). \tag{1.5}$$

The polarizability can be expanded in a Taylor series in terms of nuclear coordinates Q(t), when the driving frequency is far from electronic resonances [17].

$$\alpha(t) = \alpha_0 + \left(\frac{\partial \alpha}{\partial Q}\right)_0 Q(t) + \cdots \tag{1.6}$$

Here, α_0 and $\left(\frac{\partial \alpha}{\partial Q}\right)_0$ respectively represents the static polarizability and electronic-nuclear motion coupling coefficient. The coupling strength between the nuclear and electronic coordinates can be interpreted as the second term in Eq 1.6. A classical harmonic oscillator can be used to describe nuclear motion along Q:

$$Q(t) = 2Q_0 Cos(\omega_v t + \phi) = Q_0 [e^{i\omega_v t + i\phi} + e^{-i\omega_v t - i\phi}]. \tag{1.7}$$

where, Q_0 is the amplitude of the nuclear motion, ω_v is the nuclear resonance frequency and ϕ is phase of the nuclear mode vibration [17]. By assuming that the incoming field can be written as $E(t) = A\, e^{-i\omega_1 t} + C.C.$, then combining with Eq 1-7 leads to the dipole moment:

$$\mu(t) = \alpha_0 A e^{-i\omega_1 t} + A\left(\frac{\partial \alpha}{\partial Q}\right)_0 Q_0 \left[e^{-i(\omega_1 - \omega_v)t + i\phi} + e^{-i(\omega_1 + \omega_v)t - i\phi}\right] + C.C. \tag{1.8}$$

Three oscillation frequencies can be seen by the dipole moment equation (1.8). Rayleigh scattering is defined such that the scattered light has the same frequency as the incident light. The second term describes inelastic scattering (Raman-shifted frequencies), with terms referred to as Stokes ($\omega_1 - \omega_v$) and anti-Stokes ($\omega_1 + \omega_v$), respectively.

Note that, both Raman-shifted scattered frequencies are proportional to the coupling term $\left(\frac{\partial \alpha}{\partial Q}\right)_0$, whose the non-zero values are required along with selection rules for Raman active vibrational modes in a molecule. It is also worth mentioning that Rayleigh scattering is a coherent process whereas spontaneous Raman scattering is incoherent. The random phase (ϕ) in Eq 1.7 is why spontaneous Raman scattering is an incoherent process.

1.1.3   Classical description of coherent Raman scattering

The spontaneous Raman scattering discussed in the preceding section is usually a weak process. The scattering cross section per unit volume for Raman Stokes scattering is about 10^{-6} cm^{-1} for condensed materials. As a result, only around 1 part in a million of the incident radiation will be scattered into the Stokes frequency when propagating through one centimetre of scattering media [19-20]. The stimulated variant of the Raman scattering mechanism, on the other hand, can produce very effective scattering when excited by a strong laser beam.

We describe coherent Raman scattering via the following two steps. The molecular electron cloud first experiences oscillations as a result of two incoming fields. The oscillations of the electron cloud produce an effective force along the vibrational degree of freedom, which drives nuclear modes. Second, the driven nuclear modes are the source of a spatially coherent modulation of the refractive properties of the material [17]. The modified refractive characteristics produce sidebands that are displaced to the third light field that propagates through the material by the modulation frequency shifted [17, 20]. As a result, the presence of two incoming fields is required to form a coherent Raman signal. A classical theory to coherent Raman scattering will be addressed below.

The vibrational motion of the molecule is assumed to be explained by a damped harmonic oscillator with a resonance frequency ω_v in the classical model. The vibrational motion of two nuclei along their internuclear axis Q can be thought of as the oscillator. The two incoming beams can be written as:

$$E(t)_{1,2} = A_{1,2}e^{-i\omega_{1,2}t} + C.C. \tag{1.9}$$

where A is amplitude and $\omega_{1,2}$ are frequencies of the two incoming beams that are assumed to be much higher than the resonance frequency ω_v. Because the incoming frequencies are considerably far from vibrational resonance frequencies, the incoming field will not drive the vibrational modes. The electron cloud surrounding the nuclei, on the other hand, can adiabatically follow the incoming fields [17]. Furthermore, when the incoming fields are sufficiently strong, nonlinear electron motions can occur at combination frequencies, including the difference frequency $\Omega = \omega_1 - \omega_2$. Thus, the combined field exerts a force F(t) on the vibrational oscillator [17, 20].

$$F(t) = (\tfrac{\partial\alpha}{\partial Q})_0 [A_1 A_2{}^* e^{-i(\Omega)t} + C.C]. \tag{1.10}$$

As seen in Eq 1.10, electron motions are coupled to the nuclear motions through the $(\tfrac{\partial\alpha}{\partial Q})_0$ factor which is assumed non-vanishing here. A time-varying force which oscillates at frequency Ω (beat frequency) is introduced by the modulated electron cloud as the result of two strong incoming beams. The nuclear displacement Q can be explained by the inhomogeneous equation of motion [17]:

$$\frac{d^2 Q(t)}{dt^2} + 2\gamma\frac{dQ}{dt} + \omega_v^2 Q(t) = \frac{F(t)}{m}. \tag{1.11}$$

where, γ is damping factor, m is reduced mass of the nuclear oscillator and ω_v represents resonant frequency of the harmonic nuclear mode [17].

The solution of the second order inhomogeneous differential equation (Eq 1.11) is given by:

$$Q(t) = Q(\Omega)e^{-i\Omega t} + C.C. \tag{1.12}$$

$$Q(\omega_v) = \frac{1}{m}(\tfrac{\partial\alpha}{\partial Q})_0 \frac{A_1 A_2{}^*}{\omega_v^2 - \Omega^2 - 2i\Omega\gamma}. \tag{1.13}$$

From Eq 1.13, we can see that the nuclear mode is driven by two incident fields. The amplitude of the vibrational motion depends on both the applied incident fields and the magnitude of the coupling of the nuclear coordinate to the electronic polarizability $(\tfrac{\partial\alpha}{\partial Q})_0$. We can also see that the amplitude of the oscillatory motion reaches its maximum value when the frequency difference Ω between two incident fields approaches the vibrational resonance frequency ω_v.

The optical properties of the material are altered by the presence of driven nuclear motion. As a result, as the applied electric fields propagate through the material, their electronic polarizability will be slightly altered. [17]. The effective macroscopic polarization in the material is the sum of the all-dipole moments:

$$P(t) = N \left[\alpha_0 + \left(\frac{\partial \alpha}{\partial Q} \right)_0 Q(t) \right] \{ E_1(t) + E_2(t) \}. \tag{1.14}$$

In Eq 1.14, the terms proportional to α_0 refers to the linear polarization of the material, however the terms proportional to $\left(\frac{\partial \alpha}{\partial Q} \right)_0$ generate the third-order contribution due to the Raman effect. Using Eq 1.9 and 1.12, we can write the third-order contribution of the polarization as:

$$P_{NL} = P(\omega_{cs})e^{-i\omega_{cs}t} + P(\omega_2)e^{-i\omega_2 t} + P(\omega_1)e^{-i\omega_1 t} + P(\omega_{as})e^{-i\omega_{as}t}. \tag{1.15}$$

Where, $\omega_{cs} = 2\omega_2 - \omega_1$ and $\omega_{as} = 2\omega_1 - \omega_2$.

Thus, the nonlinear polarization contains four terms. Two represent oscillation at fundamental frequencies ω_1 and ω_2 and two new frequencies at ω_{as} and ω_{cs}. The amplitudes of the four polarization terms can be written as [17]:

$$P(\omega_{as}) = \frac{N}{m} \left(\frac{\partial \alpha}{\partial Q} \right)_0^2 \frac{1}{\omega_v^2 - \Omega^2 - 2i\Omega\gamma} A_1^2 A_2^* = 6\epsilon_0 \chi_{NL}(\Omega) A_1^2 A_2^*. \tag{1.16}$$

$$P(\omega_{cs}) = 6\epsilon_0 \chi_{NL}^*(\Omega) A_1^* A_2^2. \tag{1.17}$$

$$P(\omega_2) = 6\epsilon_0 \chi_{NL}^*(\Omega) |A_1|^2 A_2. \tag{1.18}$$

$$P(\omega_1) = 6\epsilon_0 \chi_{NL}(\Omega) |A_2|^2 A_1. \tag{1.19}$$

where:

$$\chi_{NL} = \frac{N}{6m\epsilon_0} \left(\frac{\partial \alpha}{\partial Q} \right)_0^2 \frac{1}{\omega_v^2 - \Omega^2 - 2i\Omega\gamma} \tag{1.20}$$

$\chi_{NL}(\Omega)$ is the nonlinear susceptibility which relates input electric fields to the induced polarization of the sample. The nonlinear susceptibility has two parts: (i) imaginary and (ii) real. The real part is proportional to the nonlinear refractive index whereas the imaginary part is proportational to the nonlinear absorption coefficient.

The equations 1.16 to 1.19, respectively represents Coherent Anti-Stokes Raman Scattering (CARS), Coherent Stokes Raman Scattering (CSRS), Stimulated Raman Gain (SRG) and Stimulated Raman Loss (SRL). In this thesis, I used SRL as the contrast mechanism for our hyperspectral nonlinear optical imaging. Fig. 1.2, schematically represents these types of third-order nonlinear Raman response [17], where the y axis represents the amplitude of the polarization at the Anti-Stokes frequency Eq. (1.16).

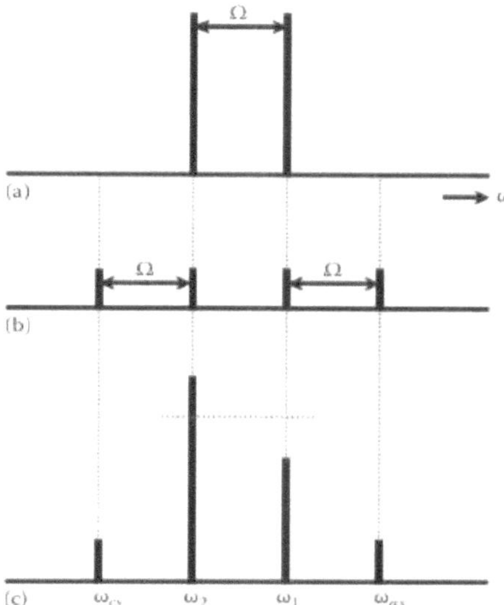

Fig. 1.2. Coherent Raman Scattering frequencies. (a): Represents incident fields at frequencies of ω_2 and ω_1. (b) Each incident frequency creates side bands by $\pm \Omega$, producing ω_{cs} and ω_1 for the ω_2 frequency, and ω_2 and ω_{as} for ω_1 frequency. (c): The intensities of the coherent Raman frequencies after passing through the material. The ω_2 frequency has experienced a gain while the ω_1 frequency experienced a loss. Although amplitudes of all four different nonlinear polarization components all depend on the magnitude of the same χ_{NL}, this does not mean the detected signals of the four coherent Raman scattering techniques are of similar strength. This figure is adapted from reference [17]. Here, the y-axis represents amplitude of the polarization.

1.1.4 CRS signal intensity

In the classical approach, an explicit assessment of Maxwell's wave equations, which tie the induced polarization to a radiating coherent field, is necessary to characterise energy flow. In a coupled wave equation technique, all of the waves ($\omega_{as}, \omega_{cs}, \omega_1, \omega_2$) involved must be considered. The coupled equations are then integrated over the (macroscopic) volume containing all the molecules to determine the energy exchange between the waves and the material as well as among the waves [17]. The solution for the nonlinear field at the anti-Stokes frequency is:

$$E_{as}{}^{(3)} = A_{as}e^{-i\omega_{as}t} + C.C. \tag{1.21}$$

The equivalent intensity for this field is calculated as follows:

$$I(\omega_{as}) = \frac{\epsilon_0 c}{2}|A_{as}|^2. \tag{1.23}$$

The anti-Stokes field is generated only by the nonlinear polarization oscillating at ω_{as} in the lowest order coherent Raman interaction. The intensity can be presented by [17]:

$$I(\omega_{as}) \propto |\chi_{NL}|^2 I_1^2 I_2. \tag{1.22}$$

Similarly for coherent Stokes we have:

$$I(\omega_{cs}) \propto |\chi_{NL}|^2 I_2^2 I_1. \tag{1.23}$$

The coherent Stokes and anti-Stokes contributions are detected as homodyne signals in the preceding description, which means that the signals are directly proportional to the modulus squared of the nonlinear polarization [17]. Because coherent Stokes and anti-Stokes have a different frequency from the incident laser frequency and smaller intensity than the incident intensity, their detection is required using PMT combined with an optical filter to remove the incident laser. Signal generation in both coherent Stokes and anti-Stokes are parametric processes, which means there is no energy exchange with the material.

In SRL, the signal is detected at frequency ω_1. The $P(\omega_1)$ is the source of the nonlinear field $E_1{}^{(3)}$ which represents the inhomogeneous part of Maxwell equations. Because the nonlinear radiation's frequency is similar to that of the primary light field E_1, we could expect to observe interference effects between two fields (scattered and incident light). A local oscillator can be viewed as the fundamental E_1 field [17]. The total intensity detected at ω_1 frequency is:

$$I(\omega_1) = \frac{\epsilon_0 c}{2}\left|E_1^{(3)} + E_1\right|^2 \propto \left|E_1^{(3)}\right|^2 + |E_1|^2 - 2I_1 I_2\, Im(\chi_{NL}). \quad (c\colon Speed\ of\ light) \qquad (1.24)$$

Because of the presence of the driven oscillator, the overall intensity at the ω_1 frequency is attenuated, as shown in Equation 1.24. Destructive interference between the induced field and the fundamental field causes the loss in the ω_1 channel. The dissipative element of the interaction, as described by the imaginary part of the nonlinear susceptibility, is responsible for the attenuation. In the ω_2 channel, the E_2 excitation field acts as the local oscillator, which can be presented as [17]:

$$I(\omega_2) \propto \left|E_2^{(3)}\right|^2 + |E_2|^2 + 2I_1 I_2\, Im(\chi_{NL}). \qquad (1.25)$$

We can observe from Equation 1.25 that the intensity at the ω_2 frequency is increasing. Constructive interference between the induced field and the driving field E_2 causes the gain in the ω_2 frequency. When modulation techniques are utilized, the heterodyne portion of the signal can be detected separately, and the resulting SRG signal is proportional to the dissipative part of the coherent Raman interaction [17].

1.2 Signal-to-Noise in SRS microscopy

In this section, we compared the SNR value in spontaneous Raman microscopy with that in CRS microscopy at a single vibrational frequency in a single image pixel. We will then discuss the importance of three different noise sources in CRM systems originating from signal detection devices as well as signal detection schemes [20].

A pixel in an image that contains important information should be differentiated from background noise. The sensitivity of an imaging system is defined as the ability to differentiate a signal from noise within a certain time period. The SNR value is typically used to quantify this capability [20]. The signal is the desired measurement's mean value (avg), which is often transformed to the detector's root mean square voltage output. The standard deviation (σ) of the measured value arising from any random fluctuations in the laser source or detector is referred to as noise [20]. The SNR can be defined as:

$$SNR = \frac{avg}{\sigma} = \frac{V_s}{\sqrt{\Sigma_i V_i^2}}. \qquad (1.26)$$

Here, V_s represents the signal voltage, and V_i denotes each type of noise voltage. An SNR greater than 1 is necessary to separate signal from noise. Longer signal integration durations can boost the SNR's value. The minimal signal-integration time and maximum imaging speed are determined by the sensitivity of an imaging system [20].

Different Raman microscopic techniques use distinct optical processes, employ different detection methodologies, and as a result, signal levels will vary. Noise sources include shot noise, laser-intensity 1/f noise, and detector Johnson noise [20].   Shot noise is produced by stochastic fluctuations in both the photocurrent (i_P) and the detector dark current (i_D) caused by the quantum nature of electrons, and takes the following form after a photodetector [21]:

$$V_{shot} = \sqrt{2e(i_P + i_D)\Delta f}.R_L. \tag{1.27}$$

Here, e is the elementary charge, Δf is the detector bandwidth, and R_L is load resistance. Usually, i_D is much smaller than i_P.

The origin of laser-intensity 1/f noise (also known as pink noise) remains under discussion; however, it can have a major impact on CRS imaging. σ_{RIN} is a frequency-dependent laser intensity 1/f noise that is commonly presented as a noise power scale [decibels relative to the carrier/frequency (dBc/Hz)], [20].

$$V_{1/f} = \sqrt{\frac{10^{\sigma_{RIN}/10}}{2\Delta t}}\, G * P * R_L. \tag{1.28}$$

where, G is the responsivity of the detector (A/W), P is average light power, R_L is load resistance and Δt represents signal collection time. Thermal perturbation of electrons in a resistor causes Johnson noise, the blackbody radiation emitted into the circuit by the resistor, and it is temperature dependent [20, 22].

$$V_J = \sqrt{4K_B T R_l \Delta f}. \tag{1.29}$$

where, K_B is the Boltzmann constant and T is the temperature in Kelvin.

1.2.1   SNR in spontaneous Raman microscopy

In most cases, spontaneous Raman microscopy necessitates measuring the complete broadband Raman spectrum. As a result, highly sensitive arrays of avalanche photodiodes or charge-coupled devices (CCDs) are commonly used to detect signals. The detector can generate a signal at its output as [20]:

$$V_{Raman} = i_{Raman}R_l = N\sigma\left(\frac{P}{A}\right)GsR_l \tag{1.30}$$

where, N is the number of molecules in the sample of interest, σ is each molecule's Raman scattering cross section at a specific wavelength, P is the input laser power, A is the exciting beam area at focus, R_l is the load resistance, G is the detector's responsivity, and s is the signal collection efficiency [20]. A typical value for a Raman cross section is on the order of 10^{-29} cm^2 [23].

Under tight-focusing conditions, we may have A = 0.09 μm^2 and s ~ 20% employing a water immersion objective lens with a numerical aperture of 1.1 for excitation-beam delivery and signal gathering. By assuming N = 10^9 and turning the laser power up to 100 mW, while G = 100 A/W, R_l = 100 kΩ we can achieve $V_{Raman} \approx 2 \times 10^{-6}$ volt.

Now, let's assume that T = 200K, and the dark current detector is negligible while the laser has a σ_{RIN} value of -110 dB/Hz (which is a typical value for some commercially available lasers), then we can calculate the contributions of each the three noise sources separately [20]:

$$V_{shot,Raman} \approx \frac{2\times10^{-10}}{\sqrt{\Delta t}}. \tag{1.31}$$

$$V_{1/f,Raman} \approx \frac{4\times10^{-12}}{\sqrt{\Delta t}}. \tag{1.32}$$

$$V_{J,Raman} \approx \frac{2\times10^{-8}}{\sqrt{\Delta t}}. \tag{1.33}$$

In this situation (dominance of Johnson noise), in order to achieve a SNR = 100, the signal integration time must be around one second. Increasing the input laser power, increasing the number of molecules in the sample, enhancing the efficiency of signal collection, lowering the detector temperature, and employing more sensitive detectors and greater load resistance are all effective approaches to speeding up signal collection. Modern Raman microscopy's imaging speed, on the other hand, is often still too slow to capture real-time dynamics in biological processes. CRS microscopy has solved this challenge by providing far greater sensitivity and imaging speed [17, 19-20].

1.2.2   SNR in SRS microscopy

Both SRL and SRG can be used as contrast mechanisms in SRS microscopy. Because the sensitivity of these two techniques are comparable (although the background signals are not), we illustrate signal detection in the SRL scheme. The wave equation can be used to calculate the evolution of the pump-beam amplitude (E_p) in the SRL process [20].

$$\frac{dE_p}{dz} = \frac{6\pi i}{\lambda_p n_p} \chi^{(3)} E_p E_s E_s^*. \tag{1.34}$$

The optical field amplitudes of the Stokes and pump beams, respectively, are E_s and E_p. The signal intensity of the pump beam SRL is determined by solving the differential Eq 1.34, which results in [20]:

$$\Delta I_p = -b_{SRS} Im(\chi^{(3)}) I_p I_s. \tag{1.35}$$

$$b_{SRS} \approx \frac{2.8 \times 10^4}{n_s n_p \lambda_p} z. \tag{1.36}$$

Here, b_{SRS} is a constant which depends on the experimental situation. For example, for $n_s = n_p = 1.3$, the pump beam tuned to $\lambda = 800$ nm, and under the tight-focusing condition $z = 1$ μm, we have $b_{SRS} \approx 2 \times 10^4$ [20]. Eq 1.37-38, represents the SRS signal voltage [20]:

$$V_{SRS} = i_{SRS} R_l = \frac{1}{2} P_{SRS} G s R_l = \frac{1}{2} |\Delta I_p| A f_{rep} \tau_{SRS} G s R_l. \tag{1.37}$$

$$V_{SRS} = \frac{1}{2} b_{SRS} Im(\chi^{(3)}) \frac{P_p P_s \tau_{SRS}}{A f_{rep} \tau_s \tau_p} G q s R_l. \tag{1.38}$$

P_p and P_s denote the average input power of pump and Stokes beams, respectively; f_{rep}, τ, and A denote the laser beam repetition rate, pulse width, and beam area at focus, respectively. G is the detecting system's responsiveness, and s is the signal gathering rate. SRS detects the tiny energy variations (modulated on the incident laser beam) carried by the strong laser beam. Directly measuring the strong laser beam could harm sensitive detectors such as PMTs. For detection of an SRS signal, a silicon photodiode is one of the best detectors with approximately $G = 0.5$ A/W which can be boosted up to 50A/W using a lock-in amplifier [20]. For a typical SRL scheme, with the optical power of both pump and Stokes set to 5 mW and pulse durations of ~ 1ps and $Im(\chi^{(3)}) = 10^{-21}$ m^2/V^2, we calculate $i_{SRS} \approx 1.7mA$ and $V_{SRS} \approx 85mV$ [20].

If the SRL of the pump beam is detected, all of the pump power appears on the detector, resulting in a lot of shot noise. The photocurrent generated by the pump beam is $i_p = P_p G_{PD} qs \approx 2.5mA$, which is significantly stronger than the photodiode dark current $i_D < 10^{-8}$ A. This gives us an output voltage before lock-in amplification of around $V_p = i_p R_l \approx 125mV$. This gives the shot-noise voltage after lock-in amplification step [20] as:

$$V_{shot,SRS} \approx \frac{10^{-7}}{\sqrt{\Delta t}}. \tag{1.39}$$

$$V_{1/f,SRS} \approx \frac{3 \times 10^{-7}}{\sqrt{\Delta t}}. \tag{1.40}$$

$$V_{J,SRS} \approx \frac{6 \times 10^{-8}}{\sqrt{\Delta t}}. \tag{1.41}$$

According to the preceding analysis, the laser shot noise and RIN are both slightly greater than the detector Johnson noise when the laser RIN power density is as low as 150 dBc/Hz (a typical value for a commercial OPO-based laser system). As a result, the total noise voltage can be determined using the following formula [20]:

$$V_{noise,SRS} = \sqrt{V_{shot,SRS}^2 + V_{1/f,SRS}^2 + V_{J,SRS}^2} \approx 3 \times \frac{10^{-7}}{\sqrt{\Delta t}}. \tag{1.42}$$

Thus, we can define $SNR_{SRS} \approx \frac{V_{SRS}}{V_{noise-SRS}} \sim 3 \times 10^5 \sqrt{\Delta t}$. Thus, the signal-integration time must be approximately around $0.1 \mu s$ to attain an SNR of 100.

CRS has substantially greater signal levels than spontaneous Raman scattering, allowing for fast vibrational imaging. In spontaneous Raman microscopy, Johnson noise restricts the SNR however, depending on the experimental conditions, all three noise sources can alter the SNR in CRS microscopy [20]. Based on the analysis in section 1.2, stimulated Raman imaging can boost signal acquisition speeds by three orders of magnitude compared to spountaneous Raman imaging.

1.3 Hyperspectral CRS microscopy

Broadband CRS systems were originally implemented using a narrowband pump and Stokes with an adjustable and variable frequency detuning between them. By consecutively tuning the pump and Stokes frequency - which results in scanning the frequency difference between pump and probe beams - different Raman bands can be probed. The sample is raster scanned either frame-by-frame or line-by-line for each Raman frequency detuning. Thus, multi-spectral images can be built by stacking the data. The stacked data makes a data cube which is called a hyperspectral image. Hyperspectral CRS refers to a group of broadband CRS systems which acquire the Raman spectrum sequentially. This not only provides the spatial distribution of sample components, it also provides the Raman spectrum per pixel. It indicates that the overall measurement time will be $N \times \tau$ plus the time it takes to tune the laser source to the next frequency detuning in order to acquire N Raman spectral points, each having an integration time constant τ. As this last argument shows, the choice of laser excitation source is critical in terms of speed and performance for these types of measurements. A perfect laser source would have these features:

1. **Low noise** (as discussed in the previous section SNR depends on RIN value of the laser source), this permits to decrease integration time for a given SNR.
2. **Broadband tunability** (helps recording a broader spectrum in order to more accurate sample classification based on Raman spectrum).
3. **Rapid tunability**, which decreases the total measurement time.

Several designs have been proposed which attempt to meet the aforementioned requirements. A common source for this setup is a mode-locked solid-state laser pumping an optical parametric oscillator (OPO) via its second harmonic. An intracavity electro-optical tunable Lyot filter ca be used to achieve rapid wavelength tuning of the OPO [24] resulting in a 115 cm^{-1} spectral coverage in $100\mu s$.

The "gold standards" for hyperspectral CRS microscopy are bulk-lasers pumping an external cavity like an OPO. Bulk lasers are expensive and have large dimensions and weight, but offer high peak outputs and minimal noise. High-quality tunable fibre lasers have resulted in a natural shift to fiber-laser-based CRS microscopy [25-30]. Transition to fibre lasers has various advantages, including device compactness, insensitivity to misalignment, and high peak powers; these help CRS microscopy systems extend from academic research labs into hospitals and analytical labs. Two crucial factors have to be taken into account specifically for SRS microscopy: first, low noise performance. As discussed above, SRS microscopy depends on modulation detection with a very small amount modulation transfer. Second, energy per bandwidth. To have high contrast images it is typically required to have to high power input lasers.

1.4 Spectral focusing

In biology and geology, nonlinear optical microscopy has become an increasingly useful tool. Ultrashort laser pulses with high peak intensity power are required (desired) for efficient signal generation for the nonlinear contrast mechanism involved. However, in CRM techniques the spectral resolution gained is limited by the accompanying wide bandwidth. A laser source with variable pulse width would be desirable due to the various Raman mode bandwidths of molecules that are commonly examined. By chirping (stretching in time) the pulses in a well-controlled manner, the spectral focusing technique allows facile optimization of the time and spectral widths of a broadband femtosecond laser source.

We achieve an effective spectral width corresponding to picosecond pulses by inducing an optimum chirp (frequency synchronization) on the pump and Stokes beams from femtosecond lasers using a spectral focusing method [31]. The chirp is often applied using glass materials with a large group velocity dispersion, such as the SF11 glass that we used in the experiments implemented in this thesis. The length of glass employed is carefully selected so that the chirp rate (change of the instantaneous central frequency with time) for the pump and Stokes beams are equal. As shown in Figure 1.3, this provides for a substantially narrower excitation Raman frequency band than for the case of unchirped beams, resulting in increased Raman spectral resolution. When the chirp rate is not the same for pump and Stokes, the excited Raman frequency band is much broader and the non-resonant signal rises relative to the resonant response. Furthermore, adjusting the relative time delay between the pump and Stokes beams permits facile and rapid tuning of the Raman spectrum. This is shown in Fig. 1.4. This technique enables rapid spectral scanning of the sample to determine the intensity of the SRS signal as a function of Raman frequency.

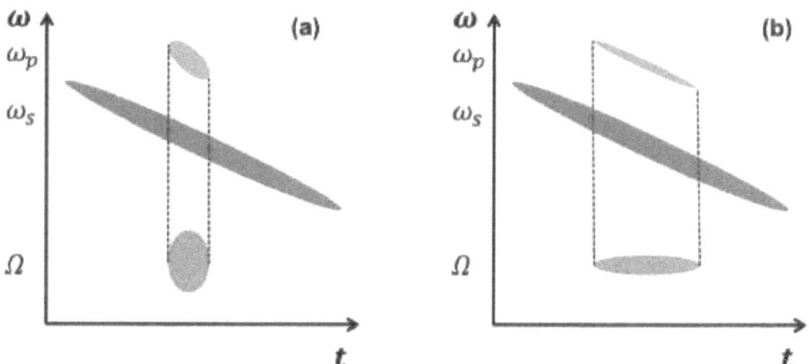

Fig. 1.3. Frequency versus time. The pump (orange) and Stokes beam (red) and the SRS signal (green) in the cases where (a) the pump beam is not optimally chirped and the spectral width of the resulting SRS signal is wider, wherein (b) the chirp is optimal and the spectrum of the resulting SRS signal is narrow.

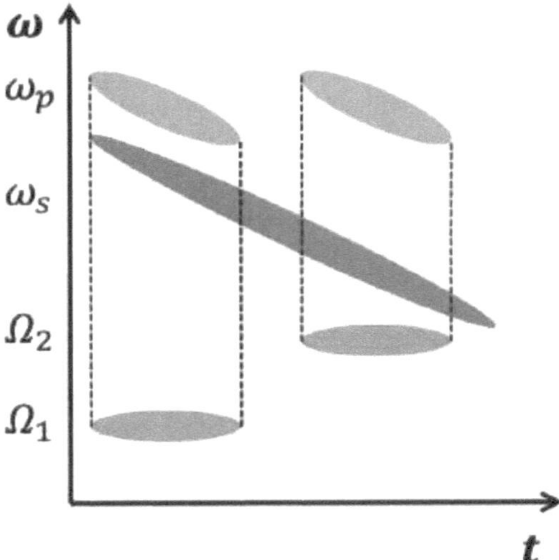

Fig. 1.4 – Frequency versus time showing the pump (orange) and Stokes (red) beams and the SRS signal (green) at two relative positions in time of the pump and Stokes beam causing a change of the center frequency of SRS signal. Changing the time delay between the Pump and Stokes permits fast tunning over Raman band of interest.

1.5 Non resonant background in SRS microscopy

The SRS signal is formed at the carrier frequency of the incident beams which operate as local oscillators and generate the signal. The Stimulated Raman Gain (SRG) and Stimulated Raman Loss (SRL) processes, which correspond to intensity gain in the Stokes beam and loss in the pump beam, respectively, are involved in the SRS process. SRS microscopy provides a spectral signal per pixel which is quite similar to the spontaneous Raman spectrum. However, there are some other nonlinear 'background' processes which can contribute to the SRS generated signal. Thus, the presence of these signals may lead to misinterpretation of the acquired signals in hyperspectral imaging. Optical processes such as cross phase modulation (XPM) [32], transient absorption [33] (TA) and photo thermal effects [34] are the most common types of aforementioned signals accompanying SRS process: these are shown in Fig 1.5. These nonlinear effects can be considered to be background signals in SRS microscopy.

Through the optical Kerr effect, the pump beam induces a change in the nonlinear refractive index at the focus in the XPM process. XPM causes changes in the propagation of the probe (Stokes) beam that can appear in the SRL signal channel. Only virtual energy levels are involved in this third-order nonlinear process, and its intensity is inversely proportional to the probe beam wavelength [35]. As a result, the XPM signal can be considered wavelength independent over a short spectral range. Because transient absorption and photothermal effects are dependent on electronic transitions, their wavelength dependence is typically slow (typically over tens of nm) for large molecules. The SRS signal, on the other hand, comes from vibrational transitions often with very sharp spectral characteristics [35]. Transient absorption (TA) is a pump-probe process in which electronic energy levels are excited. Excited state absorption, induced fluorescence, and ground state photobleaching processes are the three types of TA, with the latter two typically being indistinguishable [35]. Thermal lensing effects are caused by local heating caused by pump photon absorption, which spatially changes the temperature-dependent refractive index at the Gaussian focus, thus altering probe beam propagation [35].

These heterodyne modalities yield spectrally overlapping background signals which overlap with the SRS signal. Therefore, these could be misinterpreted as Raman resonant signals which would be misleading in any sample characterization based on Raman signal pattern recognition. Transient absorption can be minimized by employing longer excitation wavelengths [36]. The impacts of XPM and thermal lensing can be mitigated by collecting signals with a lens with a large numerical aperture [36-37]. MHz-frequency modulation can reduce photothermal effects [38-39]. These strategies, however, fail in the presence of a strong background and a weak SRS signal. Employing a broadband laser system could be helpful in interpreting the acquired signal by implementing some pattern recognition techniques such as those based on machine learning.

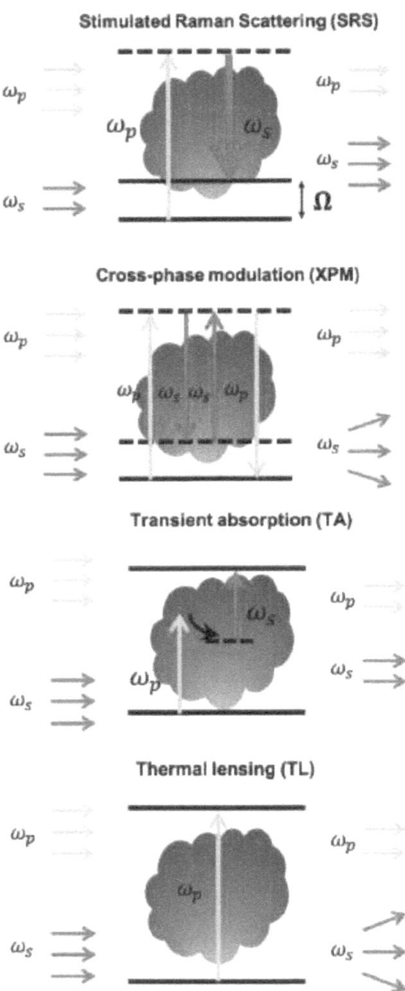

Fig. 1.5. (a): Represents SRS, (b): Cross phase modulation, (c): Transient absorption and (d) Thermal lensing effect. All the process could be interpreted as SRL which misleads the acquired Raman spectrum.

1.6 Conclusion

In this introductory chapter, we introduced nonlinear optical microscopy techniques and explored the advantages of nonlinear optical microscopy over its linear counterpart. CRS microscopy, which is a non-invasive, label-free, and non-destructive imaging approach, is introduced as a class of third-order nonlinear optical microscopy techniques. We used the classical theory of electrodynamics to introduce the spontaneous Raman scattering process in order to explain the physics underlying CRM. The fundamentals of coherent Raman scattering are then addressed. The acquisition sensitivity and signal levels of the spontaneous and coherent Raman scattering processes are also compared. It was observed that, compared to spontaneous Raman microscopy, CRM may boost signal acquisition speeds by three orders of magnitude. The combination of Hyperspectral imaging techniques and SRS microscopy results in a robust imaging modality which can determine the spatial distribution of different materials inside a sample and produce a chemical map of the sample based on a recorded Raman spectrum in each pixel. Finally, we discussed certain nonlinear and linear optical processes which generate non-resonant background signals that could be interpreted as SRS signals. Some detection and excitation approaches may help to solve the non-resonance problem. Finally, recording a broad Raman spectrum should aid in signal pattern detection.

References

[1]. J. B. Pawley, ed., Handbook of Biological Confocal Microscopy (Springer, 2006).

[2]. J.-X. Cheng, and X. S. Xie, *Coherent Raman Scattering Microscopy* (CRC Press, 2013).

[3]. J.-X. Cheng, and X. S. Xie, "Vibrational spectroscopic imaging of living systems: An emerging platform for biology and medicine," Science 350, aaa8870 (2015).

[4]. C. Zhang, D. Zhang, and J.-X. Cheng, "Coherent Raman Scattering Microscopy in Biology and Medicine," Annual Review of Biomedical Engineering 17, 415-445 (2015).

[5]. D. Polli, V. Kumar, C. M. Valensise, M. Marangoni, and G. Cerullo, "Broadband Coherent Raman Scattering Microscopy," Laser & Photonics Reviews 12, 1800020 (2018).

[6]. C. Zhang, and J.-X. Cheng, "Perspective: Coherent Raman scattering microscopy, the future is bright," APL Photonics 3, 090901 (2018).

[7]. M. C. Kao, A. F. Pegoraro, D. M. Kingston, A. Stolow, W. C. Kuo, P. H. J. Mercier, A. Gogoi, F. J. Kao, and A. Ridsdale, "Direct mineralogical imaging of economic ore and rock samples with multi-modal nonlinear optical microscopy," Scientific Reports 8 (2018).

[8]. C. H. Camp Jr, and M. T. Cicerone, "Chemically sensitive bioimaging with coherent Raman scattering," Nature Photonics 9, 295-305 (2015).

[9]. W. Denk, J. Strickler, and W. Webb, "Two-photon laser scanning fluorescence microscopy," Science 248, 73–76 (1990).

[10]. P. J. Campagnola, M.-d.Wei, A. Lewis, and L. M. Loew, "High-Resolution Nonlinear Optical Imaging of Live Cells by Second Harmonic Generation," Biophys. J. 77(6), 3341–3349 (1999).

[11]. D. Yelin and Y. Silberberg, "Laser scanning third-harmonic-generation microscopy in biology," Opt. Express 5(8), 169–175 (1999).

[12] P. Dumas, G. D. Sockalingum, and J. Sul-Suso, "Adding synchrotron radiation to infrared microspectroscopy: what's new in biomedical applications?" Trends in Biotechnology 25(1), 40 – 44 (2007).

[13] G. Turrell and J. Corset, eds., Raman Microscopy: Developments and Applications (Academic Press, 1996).

[14]. Dong Wei, Shuo Chen & Quan Liu (2015) Review of Fluorescence Suppression Techniques in Raman Spectroscopy, Applied Spectroscopy Reviews, 50:5, 387-406

[15]. A. Hopt and E. Neher, "Highly Nonlinear Photodamage in Two-Photon Fluorescence Microscopy," Biophysical Journal 80(4), 2029 – 2036 (2001).

[16]. C. V. Ramanand K. S. Krishnan. A New Type of Secondary Radiation. Nature, 121(3048):501–502, 1928.

[17]. Cheng, J. -X. & Xie, X. S. Coherent Raman Scattering Microscopy (CRC Press, 2012).

[18]. Schatz, G. C. & Ratner, M. A. Quantum Mechanics in Chemistry (Dover Publications, 2002).

[19]. Boyd, R., 2003, Nonlinear Optics, 2nd ed. Academic, New York.

[20]. C. Zhang, D. Zhang, and J.-X. Cheng, "Coherent Raman Scattering Microscopy in Biology and Medicine," Annu. Rev. Biomed. Eng. 17(1), 415–445 (2015).

[21]. Beenakker C. W. J, Büttiker M. 1992. Suppression of shot noise in metallic diffusive conductors. Phys. Rev. B 46:1889

[22]. Nyquist H. 1928. Thermal agitation of electric charge in conductors. Phys. Rev. 32:110–13

[23]. Kato Y, Takuma H. 1971. Absolute measurement of Raman-scattering cross sections of liquids. J. Opt. Soc. Am. 61:347–50

[24]. Lingjie Kong, Minbiao Ji, Gary R Holtom, Dan Fu, ChristianWFreudiger, and X Sunney Xie. Multicolor stimulated Raman scattering microscopy with a rapidly tunable optical parametric oscillator. Opt. Lett., 38(2):145–147, 2013.

[25]. Esben Ravn Andresen, Pascal Berto, and Hervé Rigneault. Stimulated Raman scattering microscopy by spectral focusing and fiber-generated soliton as Stokes pulse. Opt. Lett., 36(13):2387–2389, 2011.

[26]. Florian Tauser, Alfred Leitenstorfer, and Wolfgang Zinth. Amplified femtosecond pulses from an Er:fiber system: Nonlinear pulse shortening and self-referencing detection of the carrier-envelope phase evolution. Opt. Express, 11(6):594–600, 2003.

[27]. Thomas Gottschall, Martin Baumgartl, Aude Sagnier, Jan Rothhardt, Cesar Jauregui, Jens Limpert, and Andreas Tünnermann. Fiber-based source for multiplex-CARS microscopy based on degenerate four-wave mixing. Opt. Express, 20(11):12004–12013, 2012.

[28]. Christian W Freudiger, Wenlong Yang, Gary R Holtom, Nasser Peyghambarian, X Sunney Xie, and Khanh Q Kieu. Stimulated Raman scattering microscopy with a robust fibre laser source. Nature Photonics, 8(2):153– 159, 2014.

[29]. Claudius Riek, Claudius Kocher, Peyman Zirak, Christoph Kölbl, Peter Fimpel, Alfred Leitenstorfer, Andreas Zumbusch, and Daniele Brida. Stimulated Raman scattering microscopy by Nyquist modulation of a two branch ultrafast fibre source. Opt. Lett., 41(16):3731–3734, 2016.

[30]. Francesco Crisafi, Vikas Kumar, Antonio Perri, Marco Marangoni, Giulio Cerullo, and Dario Polli. Multimodal nonlinear microscope based on a compact fiber-format laser source. Spectrochimica Acta - Part A: Molecular and Biomolecular Spectroscopy, 188:135–140, 2018.

[31]. T. Hellerer , A. M. K. Enejder , and A. Zumbusch , " Spectral focusing: High spectral resolution spectroscopy with broad-bandwidth laser pulses ," Appl. Phys. Lett. 85 , 25 – 27 (2004)

[32]. K. Ekvall, P. van der Meulen, C. Dhollande, L. E. Berg, S. Pommeret, R. Naskrecki, and J. C. Mialocq, "Cross phase modulation artifact in liquid phase transient absorption spectroscopy," J. Appl. Phys. 87(5), 2340–2352 (2000).

[33]. D. Fu, T. Ye, T. E. Matthews, B. J. Chen, G. Yurtserver, and W. S. Warren, "High-resolution in vivo imaging of blood vessels without labeling," Opt. Lett. 32(18), 2641–2643 (2007).

[34]. K. Uchiyama, A. Hibara, H. Kimura, T. Sawada, and T. Kitamori, "Thermal lens microscope," Jpn. J. Appl. Phys. 39(9A), 5316–5322 (2000).

[35]. D. Zhang, M. N. Slipchenko, D. E. Leaird, A. M. Weiner, and J. X. Cheng, "Spectrally modulated stimulated Raman scattering imaging with an angle-to-wavelength pulse shaper," Opt. Express **21**(11), 13864–13874 (2013).

[36]. D. Zhang, M. N. Slipchenko, and J.-X. Cheng, "Highly sensitive vibrational imaging by femtosecond pulse stimulated Raman loss," J Phys Chem Lett 2(11), 1248–1253 (2011).

[37]. C. W. Freudiger, W. Min, B. G. Saar, S. Lu, G. R. Holtom, C. He, J. C. Tsai, J. X. Kang, and X. S. Xie, "Labelfree biomedical imaging with high sensitivity by stimulated Raman scattering microscopy," Science 322(5909), 1857–1861 (2008).

[38]. S. Berciaud, L. Cognet, G. A. Blab, and B. Lounis, "Photothermal heterodyne imaging of individual nonfluorescent nanoclusters and nanocrystals," Phys. Rev. Lett. 93(25), 257402 (2004).

[39]. S. Hiki, K. Mawatari, A. Hibara, M. Tokeshi, and T. Kitamori, "UV excitation thermal lens microscope for sensitive and nonlabeled detection of nonfluorescent molecules," Anal. Chem. 78(8), 2859–2863 (2006).

[40]. H. Rigneault and P. Berto, "Tutorial: Coherent Raman light matter interaction processes", APL Photonics 3, 091101 (2018).

Chapter 2

Supercontinuum generation fibre source for SRS microscopy

2. Introduction

As discussed in Chapter One, Coherent Raman Microscopy (CRM) encompasses a family of nonlinear optical imaging techniques which provide label-free, chemical-selective, 3D sectioning of samples and has found use in fields ranging from biology to medicine, to mineralogy [1-6]. Although Raman imaging based on a single vibrational resonance can be effective, non-resonant background signals and overlapping bands can reduce both contrast and chemical specificity in CRM. For target identification, tuning across the Raman spectrum is critical. For the case of modulation transfer schemes, other nonlinear optical processes, including two-photon absorption (TPA), excited-state absorption (ESA), cross phase modulation (XPM) and thermal lensing (TL), can couple (within the focal volume) the propagation of the Pump and Stokes beams and therefore appear as a modulated signal [7]. These background signals may even appear bright in an image but do not correspond to Raman resonant signals (i.e. a false positive) and therefore reduce chemical-specific contrast. Single Raman shift imaging can be misleading: only by tuning across the Raman spectrum and identifying the peaks can the nonlinear optical response be confirmed as being Raman resonant [8]. For this reason, in many applications multiple Raman bands are scanned using either multiplex or hyperspectral Raman imaging [4, 9-11]. In hyperspectral imaging, the instantaneous Pump-Stokes frequency difference must be tuned over the Raman spectral region of interest. For narrow band (picosecond) pulses, this can be achieved by tuning the center frequency of either the Pump or Stokes input lasers. Alternatively, spectral focussing of linearly chirped femtosecond (fs) pulses [12-17] can be applied, leading to rapid tunability over the femtosecond laser bandwidth by scanning the time delay between the chirped pulses. In spectral focussing CRM, however, the Raman tuning range is limited by the bandwidths of the input fs sources. For example, synchronized fs oscillators or a fs oscillator/optical parametric oscillator (OPO) combination will typically have a rapid (i.e. over the laser bandwidth) Raman tuning range of <300 cm^{-1}. To achieve broader Raman tuning ranges, the central frequency of one of the input fs lasers must be tuned, significantly reducing hyperspectral scan speeds. One approach to overcoming the bandwidth limitations of typical fs laser sources is to use supercontinuum generation in photonic crystal fibres (PCFs) in the anomalous dispersion regime, providing extremely broadband, temporally-synchronized outputs [14-16, 18]. Two serious drawbacks of PCF-based sources for CRM are that (i) they offer relatively low output power and (ii) they typically have significant source noise (i.e. modulation instabilities) in RF spectral regions important for imaging [19, 20]. For Coherent Anti-Stokes Raman Scattering (CARS) microscopy applications, this added source noise is inconvenient but not limiting [18]. For Stimulated Raman Scattering (SRS) microscopy [1, 2, 21], however, source noise can be a serious limitation. Despite the drawbacks of low power and (potentially) high source noise, PCFs have found some application in SRS microscopy [15, 16, 22]. Further enhancements using balanced detection to reduce source noise [23-25] or active gain to increase power [26] offer possible alternate routes to making the PCF sources more widely applicable, although at the cost of increased system complexity.

2.1 Chapter goals

In this chapter we first introduce spectral broadening utilizing supercontinuum generation in microstructured fibres. Different nonlinear optical processes which cause supercontinuum generation will be discussed and analyzed based on their noise properties. Later we show that All Normal Dispersion (ANDi) PCFs can generate a broad supercontinuum with minimal added noise and increase the scan range to 1000 cm^{-1} without having to tune any lasers and avoid the need for balanced detection schemes. We also demonstrate their use as a simple source for hyperspectral SRS microscopy. As a short pulse propagates in an ANDi fibre, the spectral broadening it experiences is uniquely due to self-phase modulation [27, 28], yielding a broad linearly chirped output spectrum with very small intensity fluctuations [29]. This process is highly efficient, providing high power throughput for imaging applications. Furthermore, it is a completely passive system which can be added to any standard spectral focussing SRS microscopy setup without the need of additional detectors or active optical components.

It is important to consider the various roles that the source noise spectrum can play in modulation transfer CRM [30]. There are typically two RF spectral regions of interest. One is in a band around the modulation frequency, typically in the 1-10 MHz range. In this range, source noise in the modulated (non-detected) beam is relatively unimportant: it becomes important, however, for the detected beam (vide infra). There is a second frequency range which is important for image formation in raster scanning laser microscopy. This is the noise spectral power in the range of the inverse pixel dwell time, typically in the 10-50 kHz range. The pixel dwell time acts like a low pass filter: it averages over all higher frequency noise but remains affected by lower frequency (pixel-to-pixel) noise. In modulation transfer CRM, the signal power in the detected beam depends linearly on the input power of each of the Pump and Stokes beams. This means that, although the noise in the non-detected (modulated) beam around the (MHz) modulation frequency is unimportant, pixel-to-pixel variations in the average power of the non-detected beam will lead to level variations in the detected signal averaged over the pixel dwell time. These changes in the 'quasi-DC' signal level from pixel-to-pixel lead to a noisy 'speckle-like' image, despite a high signal-to-noise ratio (at the modulation frequency) within a given pixel. As an illustration of this point, we previously demonstrated that kHz source noise can lead to detrimental 'image speckle' in an all-fibre CRM scheme [31].

In this thesis, we present a standard Stimulated Raman Loss (SRL) scheme wherein the Stokes (ANDi-based) beam is modulated, and the Pump loss signal is detected by a lock-in amplifier. In this case, the noise spectrum of the modulated Stokes is unimportant in the (MHz) modulation band. However, it can be advantageous (i.e. removes nonlinear absorption background) to use a Simulated Raman Gain (SRG) scheme [7], wherein the Pump is modulated and the gain in the Stokes is detected by a lock-in amplifier. In this implementation, the noise spectrum of the Stokes (here, ANDi-based) in the MHz range will matter greatly. Therefore, our demonstration of the low noise of the ANDi source in both the kHz (image formation) and MHz (modulation frequency) ranges will encourage its use in both SRL and SRG schemes.

2.2 Supercontinuum Generation

When a short, strong laser pulse travels through a nonlinear material, the spectrum of the propagated laser beam experiences a broadening as a result of some nonlinear optical process involved. Supercontinuum generation (SCG) is the name given to this phenomenon [32]. Alfano and Shapiro reported a spectacular case of such behaviour in 1970 [32-33]. When laser pulses of 4 ps duration tuned to 530 nm wavelength were focused into samples of calcite, quartz, sodium chloride, and numerous glasses, they found widening over much of the visible spectrum. The spectrum broadening was accompanied by self-focusing of the laser light to beam diameters of the order of 20 microns. Frok et al. observed SCG production using femtosecond pulses (1983) [1] and Corkum et al. discovered SCG in gases (1986) [34].

Although SCG has previously been shown in bulk materials and ordinary nonlinear fibres [33-37], the ability to modify dispersion parameters in photonic crystal fibre (PCF) was groundbreaking. The link between the pump wavelength and the dispersion profile of the fibre determines the bandwidth and characteristics of the generated SC spectrum. The dispersion profile design flexibility of PCF allows the fibre to be adapted to available pump sources and the SC qualities to be tailored [36, 51].

The pump source must meet stringent conditions in order to generate ultra-broadband and highly coherent spectra which keep a defined and stable phase relation between different colours in the generated spectrum. To maintain high temporal coherence in the traditional configuration using a single zero dispersion wavelength (ZDW) fibre and pumping in the anomalous dispersion regime, highly stable pulses of typically less than 50 fs duration and nano joule pulse energy are required [37]. In this situation, soliton dynamics dominates the broadening, particularly the break-up of the injected pulse due to soliton fission [38]. Due to rising noise amplification through modulation instability (MI) gain, the SC generation dynamics become particularly sensitive to fluctuations in the input pulse and pump laser shot noise over longer pulses [39,40]. As a result, these ultra-broad SCs have a complex temporal profile and phase distribution, significant fine structure across their spectral bandwidth, and pulse-to-pulse intensity and phase variations, especially if not pumped by extremely short pulses [41,42]. The SC's pulse-to-pulse intensity and phase fluctuations make it an unreliable light source for SRS microscopy where the contrast mechanism involves modulation transfer between two laser pulses as discussed in the previous chapter [43].

The remove of soliton fission in PCF with a convex and flattened dispersion profile exhibiting two closely spaced zero-dispersion wavelengths (ZDWs) centered near the pump [44] is one technique for generating coherent and re-compressible SC spectra. The resulting stable and coherent SC has been successfully applied in CARS microscopy [45] and includes two unique spectral peaks on the normal dispersion side of each ZDW [46]. However, for applications needing continuous broadband spectra, the spectral content gap between the two ZDWs is inconvenient [58].

Pumping exclusively in the normal dispersion domain avoids soliton formation, although this is frequently associated with drastically lower spectral bandwidths due to the rapid temporal broadening of the input pulse as the result of self-phase modulation [47,48]. Pumping a fibre in an anomalous dispersion regime results in more broadening but it is noisy. In contrast, pumping a fibre in an all normal dispersion regime provides less broadening but less noise even with a pump pulse duration of a few hundered femtoseconds [39]. Optimizing fibre dispersion properties may be able to overcome the limitations of this trade-off.

2.2.1 Dispersion

Properties of a generated supercontinuum within a fibre depend on the dispersion profile of the fibre and the pump wavelength laser. Pumping a fibre near its ZDW in the anomalous dispersion regime results in a greater-than-octave broadening of the input pump [49]. The material dispersion dominates in ordinary silica fibres and the ZDW which is at 1.3µm wavelength, far from the emission wavelength of the most commonly used femtosecond pump sources based on Ti:Sapphire (800 nm) or Ytterbium (1µm), is a critical issue in SCG in ordinary silica fibre. PCFs have more design degrees of freedom, allowing for customization of the dispersion profile, adapting it to certain pump sources and modifying the SC properties. The light in PCF is directed through a silica core which is surrounded in a photonic crystal cladding of air holes, most often structured in a hexagonal lattice pattern. This is shown schematically in Fig. 2.1.

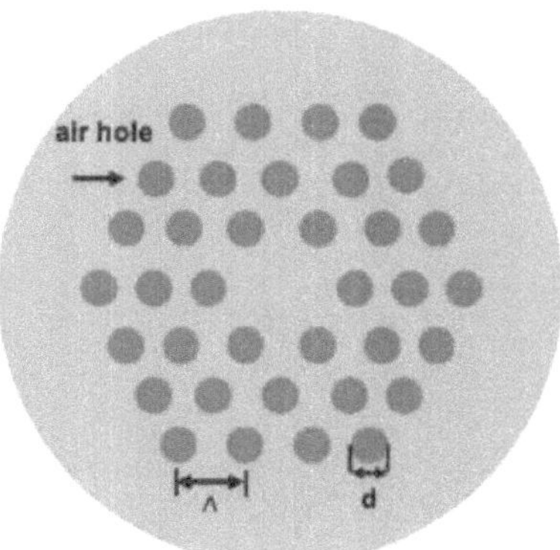

Fig. 2.1. Schematic illustration of a PCF with hexagonal lattice of air holes. Λ and d, are define as pitch and hole diameter. Altering these two results in engineering dispersion of the PCF, which material properties is also including.

The wavelength position of the ZDW can be adjusted and shifted by modifying the two design parameters pitch Λ and relative hole diameter $d/Λ$. Furthermore, the slope of the dispersion profile can be adjusted within specific limits, allowing for the design of PCF with two ZDWs, for example. Design modification flexibility is achieved by changing the dimensions of the fibre by tapering or by utilizing rings of air holes of various diameters or structures other than hexagonal.

The group velocity dispersion (GVD) dispersion parameter is defined as:

$$D = -(\frac{2\pi c}{\lambda^2})\beta_2.$$

(2.1)

where, λ is wavelength, c is speed of light in vacuum and

$$\beta_2 = \frac{\partial}{\partial\omega}\frac{1}{v_g}. \qquad (v_g \text{ is the group velocity})$$

(2.2)

The wavelength range where $\beta_2 > 0$,(D < 0) is called the normal dispersion regime, in which the group velocity decreases with wavelength. On the other hand, where $\beta_2 < 0$,(D > 0), which is called the anomalous regime group velocity behaves differently. The ZDW is where $D = 0$. The ANDi fibre used in our research made use the traditional stack-and-draw approach [61] based on pure silica tubes. Fig. 2.2(a) shows a scanning electron micrograph (SEM) of the end face of the ANDi fibre used in this thesis. The hole diameter d is set so that d/Λ = 0.39 and the distance between the holes in the cladding is $\Lambda=1.42$ µm. This fibre exhibits weak normal dispersion around the pump wavelength, as illustrated in Fig.2.2b. Fig. 2.2c shows a typical power-dependent spectrum for this ANDi fibre. In this demonstration, we used a 75.6 MHz repetition rate to launch a 140fs pulse with a wavelength of 1042 nm. The maximum input pulse energy used was 13.2 nJ, equal to 1 W input power, and the optimal input coupling efficiency was 65 percent. Fig.2.2 d shows an example output spectrum obtained with a 254 mW input pump power. The 6 dB bandwidth at this power is around 300 nm.

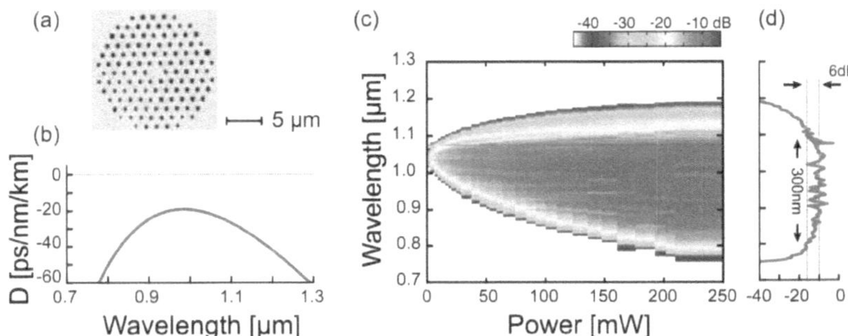

Fig. 2.2: (a) SEM of the end face of the ANDi fibre used here, (b) the calculated dispersion for this fibre, (c) the experimental power-dependent output spectrum using a ~140 fs pulse from an Yb-doped solid-state laser operating at 1042 nm, with pulse energies up to 13.2 nJ, (d) A typical output spectrum generated using 254 mW input pump power.

GVD is a linear factor which causes temporal broadening of an initially transform-limited pulse in both the normal and anomalous dispersion regimes (in the absence of nonlinearity). However, in the presence of nonlinearity, the sign of the GVD becomes a significant element, determining the effects involved in the spectral broadening dynamics [51].

2.2.2 Self-phase modulation

The intensity dependence of the refractive index causes self-phase modulation (SPM), which leads to the spectral broadening of optical pulses [32]. The time-dependent temporal intensity $I(t)$ of an optical pulse induces a modulation of the refractive index $n = n_0 + n_2 I(t)$. Neglecting the impact of dispersion and loss in the fibre, n_0 and n_2 represent the linear and nonlinear refractive indices, respectively. A time dependent phase $\phi(t)$ and, as a result, a time dependent instantaneous frequency is introduced by the intensity dependent refractive index.

$$\omega(t) = -\frac{d\phi}{dt} = \omega_0 - \gamma P_0 \frac{\partial U}{\partial t} z. \tag{2.3}$$

Here, we define intensity of the input laser as:

$$I(t) = (\frac{P_0}{A_{eff}})U(t). \tag{2.4}$$

P_0 is the peak power, A_{eff} is the effective mode field area, and $U(t)$ is normalized intensity profile [51]. In Eq 2.3, γ is called the non-linear parameter which is given by:

$$\gamma = (n_2\omega_0)/(cA_{eff}). \tag{2.5}$$

SPM generates new spectral components whose separation from ω_0 grows with nonlinearity, peak power, pulse slope, and propagation distance z. The spectrum is broadened towards red shifted lower frequencies at the leading edge of the pulse, where the slope of the rising intensity profile is positive, and towards blue shifted higher frequencies at the trailing pulse edge, where the slope of the intensity profile is negative.

When dispersion is taken into consideration, the two dispersion regimes exhibit fundamentally different behaviour. SPM with normal dispersion cause temporal and spectral broadening simultaneously. Importantly, SPM and dispersion can balance each other in the anomalous dispersion domain, resulting in the formation of solitons [51].

2.2.3 Soliton dynamics

In the anomalous dispersion domain, a soliton is generated when the nonlinear chirp caused by SPM is balanced by the linear chirp from GVD of the material [32,51]. Solitons are solutions of nonlinear Schrödinger equations, as seen in Eq.2.6. The temporal electric field envelope can be also described as Eq 2.7 [32,51].

$$i\frac{\partial A}{\partial z} + i\frac{\alpha}{2}A - \frac{\beta_2}{2}\frac{\partial^2}{\partial T^2}A + \gamma|A|^2A = 0. \tag{2.6}$$

Here, A is time-domain envelope of the input laser, α is absorption and β_2 is second order dispersion parameter. T is time measured in the travelling pulse coordinate system (frame of reference) .

$$A(t) = N \operatorname{sech}\left(\frac{t}{t_0}\right). \tag{2.7}$$

where, t_0 is a measure of the pulse duration and N is called the soliton number. The soliton number is connected to the other parameters via:

$$N^2 = L_D/L_{NL}. \tag{2.8}$$

$$L_D = \frac{t_0^2}{|\beta|}. \tag{2.9}$$

$$L_{NL} = \frac{1}{\gamma P_0}. \tag{2.10}$$

L_D and L_{NL} are the dispersive and nonlinear length scales, respectively. During propagation, the temporal and spectral features of the fundamental soliton with N = 1 remain unaltered. The spectral and temporal evolution of higher-order solitons with N > 1 is periodic [51].

2.2.4 Soliton fission

Higher order solitons' perfect periodic evolution is stable only in the absence of any perturbation. In reality, higher order dispersion and Raman scattering cause the soliton evolution to be perturbed and the injected higher order soliton breaks up into a train of fundamental solitons of the same number as the initial soliton number N [51-52]. The fundamental solitons are expelled one at a time, with the first ones having the highest amplitude and shortest lifetime [53]. This is known as soliton fission and it occurs after a distance of $L_{fiss} \approx L_D/N$ ss, when the initial soliton reaches its maximum bandwidth.

2.2.5 Soliton self-frequency shift and dispersive wave generation

Because of the soliton self-frequency shift generated by intra-pulse Raman scattering, each individual fundamental soliton undergoes a shift to longer wavelengths after soliton fission [54]. The vibrational Raman scattering of a photon by a silica unit cell results in a lower frequency photon, the self-frequency shift.

$\frac{dv_R}{dz} \propto |\beta_2|/t_0^4$ describes the dynamics of the frequency shift v_R, which leads to a stronger shift for the first ejected solitons from the fission process and, as a result, a growing separation between individual solitons as propagation distance increases [59].

2.3 SCG in anomalous regime

As can be seen from the preceding explanation, traditional (anomalous regime) SC generation in micro-structured optical fibres has a number of advantages. However, in applications where the SC's temporal profile and coherence are critical, such as in SRS microscopy, optical parametric amplification and pulse compression, some of the discussed properties become problematic and pose challenges which necessitate careful control of the nonlinear processes involved in spectral broadening dynamics.

The improved nonlinearity and improved dispersion properties of PCF enable the creation of potentially extremely coherent spectra spanning more than an octave and with pulse energies as low as 1 nJ, allowing the use of simple unamplified pump sources [36,50]. PCF's broadband single mode guiding features produce a uniform spatial profile, resulting in good spatial coherence while, filamentation effects in bulk material frequently result in more complicated spatial characteristics [36,50]. From a design point of view, the significant index difference between the silica core and the air holes results in a considerable contribution to the waveguide dispersion. The latter is very sensitive to the cladding geometry and allows the dispersion characteristics to be tailored to the needs of certain pump sources and applications [50].

In traditional SC generation, spectral broadening is inextricably linked to the soliton fission-induced breakup of the injected pulse, resulting in very complicated temporal profiles and phase distributions. As a result, suppressing soliton fission and maintaining a single ultrashort pulse in the time domain would be preferable for low noise applications. The highly structured spectra produced by soliton fission are difficult for applications which demand spectrally uniform sensitivity and signal-to-noise ratios, for example. In the development of fiber-based tunable sources based on parametric processes or pulse compression to the single-cycle regime, spectral intensity variation with wavelength is also unwanted. As a result, spectral broadening dynamics must be controlled in order to improve spectral flatness. At the end, the sensitivity of conventional SC generation to input pulse shot noise is a key issue, putting significant demands on the pump sources if highly coherent spectra are needed [36,50].

2.4 SCG in all normal dispersion regime

The supercontinuum generation in the normal GVD regime, where solitons cannot form, and spectral broadening is primarily achieved through self-phase modulation. SPM (self-phase modulation) is an internally deterministic mechanism which keeps the input pulses coherent. As a result, it is expected that creating supercontinuum in the normal GVD domain will result in excellent spectral coherence and stability. Due to the dispersive spreading of the broadened spectrum, rather than compression as in anomalous-dispersion fibre, the output bandwidth cannot be expected to compete with that of an anomalous dispersion fibre. To increase the spectral bandwidth, proper dispersion engineering is required. Confining light to a certain spectral region, on the other hand, is advantageous in applications which demand a high spectral power density, such as CRM. PCF's design flexibility permits engineering of the dispersion profile which is desirable in SC generation. One way to generate supercontinuum in the normal GVD regime is to pump a conventional fibre far below the ZDW, so that the resulting spectrum does not extend into the anomalous dispersion region. However, due to the high value of dispersion far from the ZDW, this would necessitate high power or very short pulses to overcome the short effective interaction length. A more effective approach, which was used in this thesis, is to create supercontinuum in a photonic crystal fibre with an all-normal GVD profile and low dispersion at the pump wavelength. PCF with all-normal dispersion profiles can be made over a short range of pitch (hole-to-hole distance) and hole diameter (Fig. 2.2).

2.5 Experimental setup for RF spectral noise characterization

A schematic of the spectrally resolved noise measurement setup is shown in Fig. 2.3. All measurements were performed with a femtosecond dual output laser system (InSight DS +, Spectra-Physics, USA) which produced two synchronized pulse trains at 80 MHz. The fixed wavelength output centered at 1040 nm has a transform-limited pulse duration of 220 fs with an average power of 1 W, whereas the second output was tunable over the 680-1300 nm range with a transform-limited pulse duration of approximately 180 fs and 1.5 W of average power. The 1040 nm output was sent through a variable beam splitter constructed from a half wave plate (HWP), (AHWP05M-980, Thorlabs, USA) and a polarizing beam splitter (PBS), (PBS102, Thorlabs, USA) before being passed through a Pockels cell (350-160, Conoptics, USA) used to impart a 1 MHz amplitude modulation. Another HWP was used to match the pump laser polarization to that of the fibre axis before it was focussed by an aspheric lens (C230TMD-B, Thorlabs) into the 10 cm long ANDi fibre (core diameter of 2.2 μm) shown in Fig. 2.2 (a). In fact, spectral broadening happens in just a few millimetres of fibre. The reason why we selected 10 cm fibre length was the convenience of coupling light into it and our preference for chirped pulses at the output. The ANDi fibre alignment, in terms of spatial overlap and polarization axis, was optimized for maximum spectral broadening, (56-59% coupling efficiency). The ANDi fibre output was collimated by another aspheric lens (C340TMD-B, Thorlabs) and spectrally dispersed by a prism followed by an adjustable slit to select a narrow spectral band from the ANDi supercontinuum output. To allow for comparative noise measurements, the average power within each spectral band was kept constant using a variable neutral density (ND) filter (NDC-50C-2-B, Thorlabs). The central wavelength and bandwidth of each spectral band was determined using an optical spectrum analyzer (Ando AQ6315E). The pulse train of the selected band was measured by a Si photodiode (DET10A Thorlabs), the output of which was sent to a Zurich Instruments ultra-high-frequency lock-in (UHFLI) amplifier. The sweeper function (Amplitude Noise) of the UHFLI was used for all RF spectral noise measurements reported here.

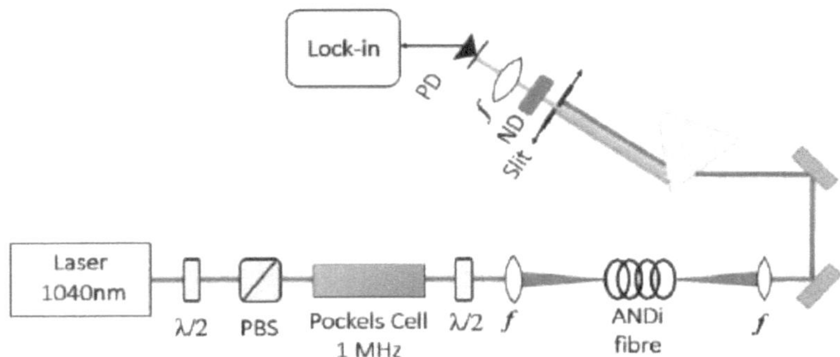

Fig. 2.3. A depiction of the spectrally resolved Noise Power Spectral Density (NPSD) measurement setup. This arrangement permits direct measurement of the NPSD in the RF (kHz-MHz) spectral domain within a selected optical bandpass (10nm) of the ANDi fibre supercontinuum output. A neutral density filter (ND) ensured that the NPSD measurements were done at constant input optical power. Illustrated are the pump laser at 1040nm, half wave plate ($\lambda/2$), polarizing beam splitter (PBS), Pockels Cell (1MHz), coupling/collimating lens (f), ANDi fibre, adjustable optical bandpass slit, neutral density filter (ND), Si photodiode (PD), and lock in amplifier.

2.6   RF Noise Power Spectral Density

In SRS microscopy, which is typically based upon modulation transfer detection combined with a raster scanning technique for image generation, source noise can limit both (i) the Signal-to-Noise Ratio (SNR) at the modulation frequency, and (ii) the pixel-to-pixel noise in the image. As discussed in the Introduction, the Stokes (here ANDi-based) source noise at the MHz modulation frequency is critical only when the Stimulated Raman Gain scheme is implemented. In contrast, the source noise at the inverse pixel dwell time (kHz) is always important in the image formation process.

Here we implemented a Stimulated Raman Loss (SRL) scheme wherein the ANDi-based Stokes is modulated, and the loss of tunable Pump is detected by the lock-in amplifier. In this case, the noise spectrum of the modulated (ANDi) Stokes is unimportant in the (MHz) modulation band. However, for Simulated Raman Gain (SRG), the noise spectrum of the ANDi-based Stokes will matter greatly. Therefore, to quantitatively determine the suitability of an ANDi fibre source for SRS microscopy, we characterized the noise of both the input laser and the ANDi supercontinuum output in RF regions around 1 MHz (a typical modulation frequency) and around 40 kHz (a typical inverse pixel dwell time) by measuring the RF noise power spectral density (NPSD in $dBV/\sqrt{Hz}$). To ensure that the noise characteristics were as would be for SRS microscopy, the input laser was amplitude modulated with a 50% duty cycle at 1 MHz using a Pockels cell. For all noise measurements reported here, we used (following the Pockels cell modulator) 500 mW average power (12.5 nJ/pulse) in the 1040 nm pump beam. The measured NPSD in the 1 MHz range are shown in Fig. 2.4 for: the electronic noise floor; the input laser; and several different 10 nm spectral bands from the ANDi supercontinuum output. The electronic noise floor was measured with all input optical signals blocked. The average optical power at the Si photodiode within each 10 nm spectral band, and for the input pump laser, were set equal by adjusting the variable ND filter. To characterize the increase in noise due to supercontinuum generation, we report the "excess NPSD" which we define to be the average increase in NPSD relative to the electronic noise floor. In the 1 MHz range (0.9-1.1 MHz), we report the average NPSD by excluding the large Pockels Cell modulation peak at 1 MHz. Around 40 kHz (20-60 kHz), we report the average NPSD over the entire range. In Fig. 2.5, we show the excess NPSD for the different supercontinuum 10 nm-wide spectral bands (centered at 940, 962, 973, 1003, and 1020 nm) as well as that of the input laser (1040 nm) before transmission through the fibre. As expected, the input laser shows little excess NPSD at 1 MHz: 0.9 $dBV/\sqrt{Hz}$ above background. However, at 40 kHz the pump laser does show a measurable 1.7 $dBV/\sqrt{Hz}$ increase in NPSD. As seen in Fig. 2.5, all supercontinuum 10 nm spectral bands show increased excess NPSD relative to the input laser: these appear similar in both RF frequency ranges, varying from 3.7-14.2 $dBV/\sqrt{Hz}$. However, the observed increases in excess NPSD are relatively small, suggesting that an ANDi source may be suitable for broadband SRS microscopy.

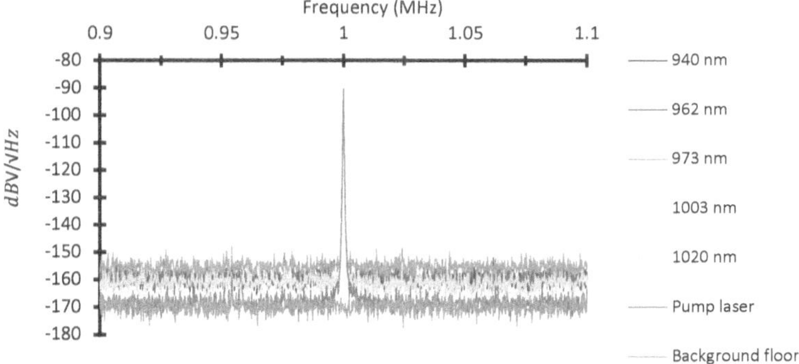

Fig. 2.4. RF Noise Power Spectral Density over [0.9-1.1] MHz at constant input optical power, shown for the electronic noise floor (gray, zero input), the 1040 nm laser (red), and as a function of 10 nm spectral band central wavelength.

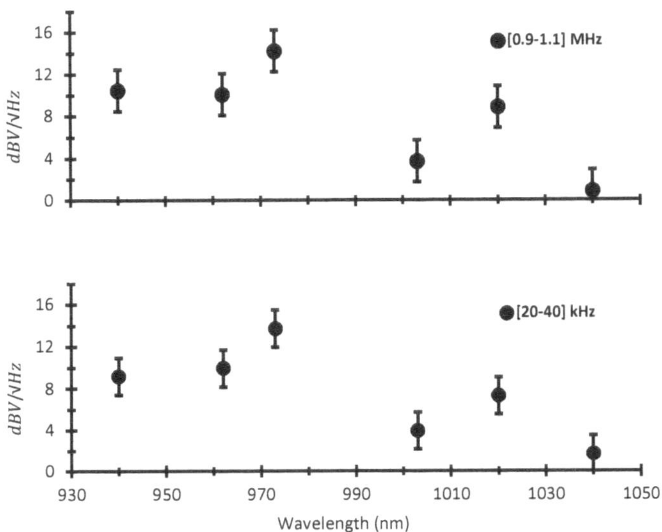

Fig. 2.5. Excess Noise Power Spectral Density relative to the electronic noise floor. Shown are the Excess NPSD as a function of 10 nm spectral band central wavelength over (a) the 0.9-1.1 MHz range, (b) 20-40 kHz range.

2.7　Hyperspectral Raman Scattering microscopy

In order to demonstrate the use of ANDi fibres for hyperspectral SRS microscopy, we used a broadband SRS spectral focusing arrangement, illustrated in Fig. 2.6. The InSight DS+ laser system provided a tunable output beam and a fixed wavelength beam (1040nm), the latter was used for supercontinuum generation in the ANDi fibre. The average power of each output was adjusted using a half-wave plate, (AHWP05M-980, Thorlabs, USA) and a polarizing beam splitter (PBS102, Thorlabs, USA). The input to the ANDi fibre was identical to that used for the spectrally-resolved noise measurements. After collimation, the fibre output was spectrally filtered (980 nm RazorEdge LP, Semrock) so as to remove all supercontinuum wavelengths below 980 nm. The supercontinuum output (acting here as the SRS Stokes beam) and the tunable InSight output (acting here as the SRS pump beam) were recombined on a dichroic mirror (DM), (DMLP 950, Thorlabs, USA). Both beams were linearly chirped by 60 cm of SF11 glass. This chirped pulse arrangement permits rapid tuning of the Raman resonant frequency by time-scanned spectral focusing. The time delay between the tunable pump and the supercontinuum Stokes beam (which tunes the instantaneous Raman frequency) was controlled by a retro-reflector mounted on a translation stage in the pump beam path. The recombined beams were sent to an inverted microscope (IX-71, Olympus, Japan) and focused into the sample with a near IR microscope objective (UPlanSapo, 20x, NA 0.75, Olympus, Japan). Galvanometer mirrors permitted raster scanning across the sample, thus providing an image. After the sample, the forward propagating beams were collected by a second objective acting as the condenser (LUMPlanFI/IR, 40x, NA 0.8 water immersion, Olympus, Japan). A function generator (DS345, Stanford Research Systems, USA) provided the 1 MHz modulation signal for the Pockels cell. The modulation reference signal was sent to a lock-in amplifier (UHFLI, Zurich Instruments). The amplitude modulated supercontinuum beam was blocked by optical filters (FES0950 SP Thorlabs, FF01-94C/SP-25 Semrock) and the transmitted pump beam was recorded using a large-area photodiode (FDS10X10, Thorlabs, USA). Typically, a few mW of optical power impinged onto the photodiode which was reverse biased at 50 V. The photodiode output electrical signal was filtered by an RF bandpass filter (#3128, KR Electronics) centred at 1 MHz frequency. The filtered signal was amplified by a transimpedance amplifier (DHPCA-100, Femto Messtechnik GmbH, Germany), providing the signal input to the lock-in amplifier. A lock-in time constant of 20 μs was used and the relative phase of the lock-in amplifier was adjusted for maximum SRS signal. Data collection and galvo synchronization was performed using ScanImage [62].

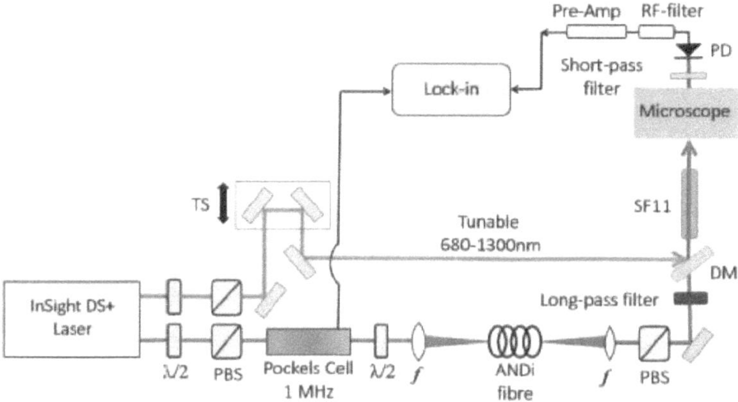

Fig. 2.6: The broadband Hyperspectral Stimulated Raman Scattering (SRS) microscopy optical arrangement using the ANDi fibre source. In this experiment, the tunable output acted as the SRS pump, whereas the ANDi fibre output acted as the broadband SRS Stokes. Depicted are the: dual output laser system, half wave plate (λ/2), polarizing beam splitter (PBS), Pockels cell (1MHz), translation stage (TS), focussing/collimating aspheric lens (f), All normal dispersion fibre as supercontinuum source (ANDi), dichroic mirror (DM), highly dispersive piece of glass (SF-11), RF-filter: electronic bandpass filter at 1 MHz central frequency, and photodiode (PD).

In order to normalize the recorded SRS spectra, we needed to determine the wavelength variation of optical power within the ANDi supercontinuum output. This can be routinely achieved by recording the borad band sum-frequency of the instantaneous pump and Stokes frequencies. In the epi-direction, a dichroic mirror (720DCXXR, Chroma, USA) directed back-reflected sum frequency generation (SFG) signals through a short pass filter (750SP, Chroma, USA) in front of a photomultiplier tube (Hamamatsu H10723-01). Samples of KDP powder (data not shown) were used to generate the broadband SFG signal. Since both SFG and SRS are linear in the pump and Stokes powers, SFG can be used to normalize (as a function of Raman shift) the spectrally resolved SRS signals from samples.

2.7.1   Fibre output supercontinuum stability

Implementing broadband hyperspectral SRS imaging requires a broadband laser source and therefore the laser source stability should also be taken into account. Small pointing changes of the beam which pumps the ANDi fibre results in drifting of the output supercontinuum spectrum. In such a case, we would not be able to do an accurate calibration of the Raman spectrum. In addition, pump laser pointing unstability could damage the fibre's facet when the pump power is high ($\sim$ 1W). Another issue which makes the output supercontinuum unstable is absorption in humid air . The small microstructures in the fibre (Fig. 2.2 a) draw the humid air inside. Humid air absorption changes not only the coupling efficiency in long-term use of the fibre (2-3 months) but also changes the spectral broadening as the result of dispersion modification inside the cores. We therefore isolated the fibre coupling setup from the rest of the lab by building a simple dry box to solve these issues. The fibre coupling box was purged with a slow flow of dry nitrogen, as shown in Fig. 2.7. The lifetime of the fibre facet increased from 1 hour to $\sim$ 2 hours when pumped with 1 W of 1040nm light.

Fig. 2.7: Stabilizing the coupling by isolating coupling setup from the lab. Dry Nitrogen filled to the box to prevent humid air absorption.

2.8 Hyperspectral SRS microscopy using ANDi fibres

In order to demonstrate the use of ANDi fibres for broadband Hyperspectral SRS microscopy, we chose two samples: neat Dimethyl Sulfoxide (DMSO) and acetaminophen powder. DMSO has two well-known Raman vibrational resonances at 2997 cm^{-1} and 2919 cm^{-1}. To obtain an SRS spectrum, the pump beam was tuned to 817 nm and both the pump and supercontinuum were adjusted to 50 mW average power at the microscope input. The lock-in time constant was set to 20µs and the pixel dwell time was 32 µs. The SRS spectrum was recorded over 94 frames with a spectral scan speed of ~ 3 cm^{-1}/s and each image took ~ 2.13 seconds. An SRS spectrum of DMSO thus obtained, along with the spontaneous Raman spectrum of DMSO, is shown in Fig. 2.8. The lower energy peak of DMSO (2997 cm^{-1}) was used for normalization to facilitate comparisons. We estimate the spectral resolution to be ~ 37 cm^{-1}, limited here by poor chirp matching [14, 63]. This can be significantly improved by carefully matching the linear and higher order chirp parameters of the pump and Stokes beams. We note, however, the current resolution is sufficient for many applications.

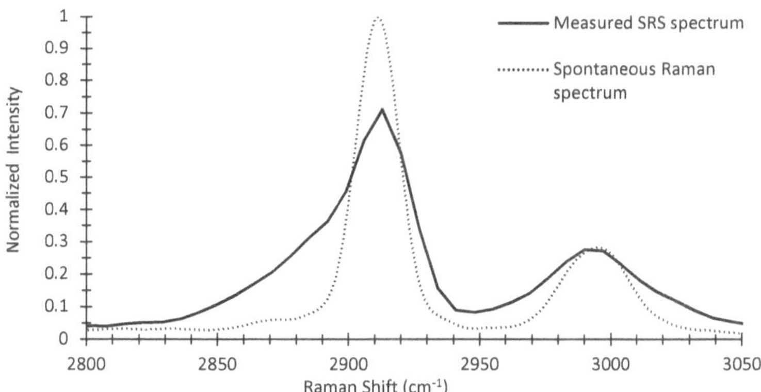

Fig. 2.8. Raman spectrum of DMSO recorded using SRS (solid line) and spontaneous Raman spectroscopy (dashed line) with the two main peaks in this region indicated. This demonstrates the utility of ANDi fibre sources for SRS spectroscopy. Normalization to peak height is based on the 2997 cm^{-1} peak.

In order to demonstrate the broadband hyperspectral SRS microscopy imaging capability of this ANDi light source, we imaged acetaminophen powder which has many Raman resonances in the fingerprint region. In this case, the pump beam was tuned to 910 nm (50 mW average power) whereas the broadband supercontinuum (Stokes beam) had 60 mW of average power, measured at the microscope input. The data acquisition scheme was the same as was used for the DMSO experiment above. The SRS spectral scan was acquired over 259 frames with a spectral scan speed of 6.85 cm^{-1}/s and each image took 2.13s. Hyperspectral SRS imaging of acetaminophen powder was recorded in the forward direction and normalized by an independently recorded SFG spectrum to correct for spectral power variation in the Stokes spectrum. At 500 mW ANDi input power (12.5 nJ/pulse, 50% duty cycle), the output Stokes bandwidth supported a continuous SRS tuning range of 858 $-1648\ cm^{-1}$. Importantly, this broad tuning range was achieved without tuning any input laser sources. A typical image is shown in Fig. 2.9 (a). As discussed above, due to the limited applied chirp and the resulting mismatch of the pump and Stokes chirps, the Raman spectral resolution in this proof-of-concept demonstration is unoptimized. Corrections to the linear and the higher order chirp terms can lead to considerably improved Raman spectral resolution [63]. We emphasize, however, that this is not a limitation of the ANDi fibre source itself. The individual microcrystallites of acetaminophen were randomly oriented within the sample and therefore the relative intensities of the Raman peaks, which are in general orientation-dependent, is not meaningful here.

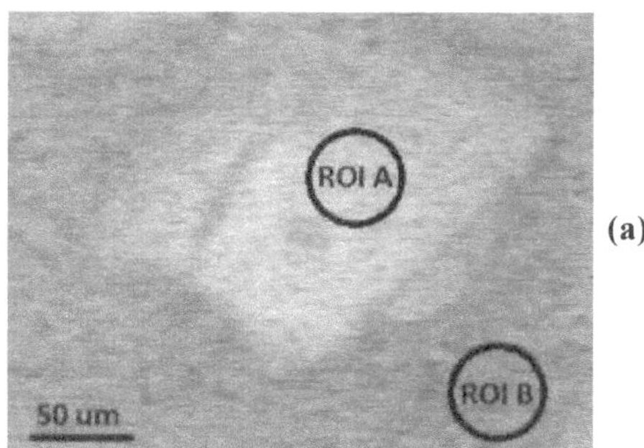

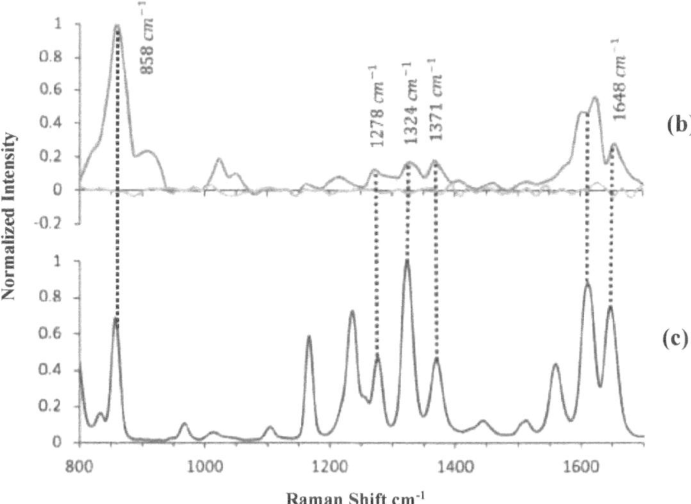

Fig. 2.9. (a) Normalized Hyperspectral SRS Microscopy image of a single acetaminophen micro-crystal. (b) The SRS Raman spectra of two Regions of Interest (ROI) are shown for the acetaminophen crystal (A, Red) and the background (B, Blue). (c) Spontaneous Raman spectrum of acetaminophen. A few of the peaks in the acetaminophen Raman spectrum are labelled in both (b) and (c), as indicated by the dashed lines.

Finally, in order to demonstrate the power of ANDi-based Hyperspectral SRS Microscopy for label-free chemical-specific image contrast, we applied spectral cross correlation of the acetaminophen Raman spectrum to the whole image shown in Fig. 2.9 (a): the spectral cross-correlation will maximize when a given pixel contains Raman peaks which match those of acetaminophen. Using spectral pattern recognition, we sought pixels which contain the set of Raman peaks (not a single peak) assigned to the target species. This spectral cross-correlation was used to generate enhanced image contrast in a molecule-selective manner: it is based on the known Raman spectrum of the target species in the spectral band of interest. The processed image is shown in Fig. 2.10. The contrast with respect to the background is much improved. We note that, in this case, the spectral cross correlation acts largely as a non-resonant background subtraction method. However, more generally, the power of the spectral cross-correlation approach is that it permits image contrast based on small changes in the Raman spectrum rather than in a single peak, such as those due to changes in the local chemical environment.

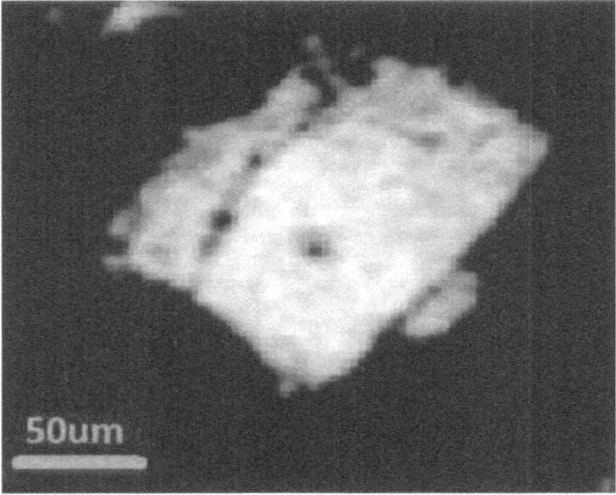

Fig. 2.10. Spectral cross-correlation of the image from Fig. 2.9 (a), with the acetaminophen Raman spectrum shown in Fig. 2.9 (b). The cross-correlation maximizes when a given pixel contains a Raman spectrum with peaks matching those of Fig. 2.9 (b). The image contrast with respect to background greatly increases.

2.9 Conclusion

In hyperspectral SRS microscopy, the ability to rapidly scan over a wide range of Raman shifts is restricted because of the limited bandwidth of the laser sources available. We have shown that, using an ANDi fibre, it is possible to greatly increase the bandwidth of one of the laser pulses without any significant increase in source noise which would otherwise have a deleterious effect on SRS performance. Using this setup, we demonstrated hyperspectral Raman imaging over a 800 cm^{-1} scan range without tuning either laser: in spectral focussing, the Raman spectrum is scanned simply by changing the time delay between pulses.

Several improvements could be made to our current proof-of-concept setup in order to further enhance its utility. The Raman spectral resolution in spectral focussing SRS is determined by the degree of linear chirp and the matching of the Pump and Stokes chirp rates. We and others have previously demonstrated that spectral focussing SRS can have spectral resolution on the order of 10 cm^{-1} via optimization of the chirp rate. The ability to optimize the chirp rate is sometimes limited by higher order dispersion; we find that the ANDi continuum output has minimal higher order dispersion based on our measurements of broad Raman spectra (i.e. the conversion of pulse time delay to wavenumber is fairly linear). The current spectral scan range, >800 cm^{-1}, could also be increased. The output of the ANDi fibre on the blue side extends to <930 nm with reasonable output power and low noise but our current implementation truncates this spectrum due available optical filters. Extending the Stokes spectrum to below 930 nm would further extend the spectral scan range of the ANDi fibre source.

Alternatively, rather than increasing the scan range, the spectral energy density of the continuum (i.e. the power per unit bandwidth) can be increased at the expense of total bandwidth. For continuum generation in ANDi fibres, the total bandwidth depends on the peak power, not the input pulse duration [64]. In the proof-of-concept demonstration presented here, we operated at maximum broadening which presents the challenge that the degree of noise often increases with nonlinearity. This demonstrated that hyperspectral SRS imaging is feasible even with the increased noise at maximal broadening. Alternatively, one could operate with less spectral broadening simply by applying a positive chirp to the ANDi fibre input pump pulse. This will increase the power per unit bandwidth in the continuum and concomitantly decrease the NPSD, allowing for higher signal-to-noise ratio imaging across a narrower bandwidth of interest. The ANDi fibre operating conditions may be readily optimized for different applications without any change in hardware. In a future research direction, we will investigate a new type of polarization-maintaining ANDi fibre which we anticipate will have much improved power spectral density in its output. In conclusion, we have characterized the Noise Spectral Power Density of an ANDi fibre and demonstrated its utility as a simple, flexible and readily implemented source for broadband hyperspectral SRS microscopy.

References

[1]. J.-X. Cheng, and X. S. Xie, *Coherent Raman Scattering Microscopy* (CRC Press, 2013).

[2]. J.-X. Cheng, and X. S. Xie, "Vibrational spectroscopic imaging of living systems: An emerging platform for biology and medicine," Science **350**, aaa8870 (2015).

[3]. C. Zhang, D. Zhang, and J.-X. Cheng, "Coherent Raman Scattering Microscopy in Biology and Medicine," Annual Review of Biomedical Engineering **17**, 415-445 (2015).

[4]. D. Polli, V. Kumar, C. M. Valensise, M. Marangoni, and G. Cerullo, "Broadband Coherent Raman Scattering Microscopy," Laser & Photonics Reviews **12**, 1800020 (2018).

[5]. C. Zhang, and J.-X. Cheng, "Perspective: Coherent Raman scattering microscopy, the future is bright," APL Photonics **3**, 090901 (2018).

[6]. M. C. Kao, A. F. Pegoraro, D. M. Kingston, A. Stolow, W. C. Kuo, P. H. J. Mercier, A. Gogoi, F. J. Kao, and A. Ridsdale, "Direct mineralogical imaging of economic ore and rock samples with multi-modal nonlinear optical microscopy," Scientific Reports **8** (2018).

[7]. P. Berto, E. R. Andresen, and H. Rigneault, "Background-free stimulated Raman spectroscopy and microscopy," Phys. Rev. Lett. **112**(5), 053905 (2014).

[8]. M. A. Houle, R. C. Burruss, A. Ridsdale, D. J. Moffatt, F. Légaré, and A. Stolow, "Rapid 3d chemical-specific imaging of minerals using stimulated raman scattering microscopy," J. Raman Spectrosc. **48**(5), 726–735 (2017).

[9]. C. H. Camp Jr, and M. T. Cicerone, "Chemically sensitive bioimaging with coherent Raman scattering," Nature Photonics **9**, 295-305 (2015).

[10]. T. W. Kee, and M. T. Cicerone, "Simple approach to one-laser, broadband coherent anti-Stokes Raman scattering microscopy," Opt Lett **29**, 2701-2703 (2004).

[11]. H. Kano, and H.-o. Hamaguchi, "Ultrabroadband (>2500cm−1) multiplex coherent anti-Stokes Raman scattering microspectroscopy using a supercontinuum generated from a photonic crystal fiber," Applied Physics Letters **86**, 121113 (2005).

[12]. T. Hellerer, A. M. K. Enejder, and A. Zumbusch, "Spectral focusing: High spectral resolution spectroscopy with broad-bandwidth laser pulses," Applied Physics Letters **85**, 25-27 (2004).

[13]. I. Rocha-Mendoza, W. Langbein, and P. Borri, "Coherent anti-Stokes Raman microspectroscopy using spectral focusing with glass dispersion," Applied Physics Letters **93**, 201103 (2008).

[14]. A. F. Pegoraro, A. Ridsdale, D. J. Moffatt, Y. Jia, J. P. Pezacki, and A. Stolow, "Optimally chirped multimodal CARS microscopy based on a single Ti:sapphire oscillator," Opt. Express **17**, 2984-2996 (2009).

[15]. E. R. Andresen, P. Berto, and H. Rigneault, "Stimulated Raman scattering microscopy by spectral focusing and fiber-generated soliton as Stokes pulse," Opt Lett **36**, 2387-2389 (2011).

[16]. H. T. Beier, G. D. Noojin, and B. A. Rockwell, "Stimulated Raman scattering using a single femtosecond oscillator with flexibility for imaging and spectral applications," Opt. Express **19**, 18885-18892 (2011).

[17]. D. Fu, G. Holtom, C. Freudiger, X. Zhang, and X. S. Xie, "Hyperspectral Imaging with Stimulated Raman Scattering by Chirped Femtosecond Lasers," The Journal of Physical Chemistry B **117**, 4634-4640 (2013).

[18]. J. G. Porquez, R. A. Cole, J. T. Tabarangao, and A. D. Slepkov, "Brighter CARS hypermicroscopy via "spectral surfing"of a Stokes supercontinuum," Opt Lett **42**, 2255-2258 (2017).

[19]. F. Lu, and W. H. Knox, "Generation of a broadband continuum with high spectral coherence in tapered single-mode optical fibers," Opt. Express **12**, 347-353 (2004).

[20]. A. M. Zheltikov, Physics-Uspekhi **49**, 605 (2006).

[21]. C. W. Freudiger, W. Min, B. G. Saar, S. Lu, G. R. Holtom, C. He, J. C. Tsai, J. X. Kang, and X. S. Xie, "Label-Free Biomedical Imaging with High Sensitivity by Stimulated Raman Scattering Microscopy," Science **322**, 1857-1861 (2008).

[22]. K. Seto, Y. Okuda, E. Tokunaga, and T. Kobayashi, "Development of a multiplex stimulated Raman microscope for spectral imaging through multi-channel lock-indetection," Rev Sci Instrum **84**, 083705 (2013).

[23]. K. Nose, Y. Ozeki, T. Kishi, K. Sumimura, N. Nishizawa, K. Fukui, Y. Kanematsu, and K. Itoh, "Sensitivity enhancement of fiber-laser-based stimulated Raman scattering microscopy by collinear balanced detection technique," Opt. Express **20**, 13958-13965 (2012).

[24]. A. Gambetta, V. Kumar, G. Grancini, D. Polli, R. Ramponi, G. Cerullo, and M. Marangoni, "Fiber-format stimulated-Raman-scattering microscopy from a single laser oscillator," Opt Lett **35**, 226-228 (2010).

[25]. C. W. Freudiger, W. Yang, G. R. Holtom, N. Peyghambarian, X. S. Xie, and K. Q. Kieu, "Stimulated Raman scattering microscopy with a robust fibre laser source," Nature Photonics **8**, 153-159 (2014).

[26]. B. Figueroa, W. Fu, T. Nguyen, K. Shin, B. Manifold, F. Wise, and D. Fu, "Broadband hyperspectral stimulated Raman scattering microscopy with a parabolic fibre amplifier source," Biomed. Opt. Express **9**, 6116-6131 (2018).

[27]. G. Agrawal, *Nonlinear Fibre Optics* (Academic Press, 2013).

[28]. A. M. Heidt, A. Hartung, G. W. Bosman, P. Krok, E. G. Rohwer, H. Schwoerer, and H. Bartelt, "Coherent octave spanning near-infrared and visible supercontinuum generation in all-normal dispersion photonic crystal fibers," Opt. Express **19**, 3775-3787 (2011).

[29]. A. M. Heidt, "Pulse preserving flat-top supercontinuum generation in all-normal dispersion photonic crystal fibers," J. Opt. Soc. Am. B **27**, 550-559 (2010).

[30]. X. Audier, S. Heuke, P. Volz, I. Rimke, H. Rigneault "Noise in stimulated Raman scattering measurement: From basics to practice." APL Photonics 5, 011101 (2020).

[31]. A. F. Pegoraro, A. Ridsdale, D. J. Moffatt, J. P. Pezacki, B. K. Thomas, L. Fu, L. Dong, M. E. Fermann, and A. Stolow, "All-fiber CARS microscopy of live cells," Opt. Express **17**(23), 20700–20706 (2009).

[32]. Boyd, R., 2003, Nonlinear Optics, 2nd ed. Academic, New York.

[33]. R. R. Alfano and S. L. Shapiro, "Emission in the region 4000 to 7000 Å via four photon coupling in glass," Phys. Rev. Lett. 24, 584–587 (1970).

[34] P. B. Corkum, Claude Rolland, and T. Srinivasan-Rao, Phys. Rev. Lett. 57, 2268

[35]. C. Lin and R. Stolen, "New nanosecond continuum for excited-state spectroscopy," Appl. Phys. Lett. 28, 216–218 (1976).

[36]. R. Alfano, The supercontinuum laser source (Springer, New York, 2006).

[37]. J. M. Dudley and S. Coen, "Coherence properties of supercontinuum spectra generated in photonic crystal and tapered optical fibers," Opt. Lett. 27, 1180–1182 (2002).

[38]. J. Herrmann, U. Griebner, N. Zhavoronkov, A. Husakou, D. Nickel, J. Knight, W. Wadsworth, P. S. J. Russell, and G. Korn, "Experimental evidence for supercontinuum generation by fission of higher-order solitons in photonic fibers," Phys. Rev. Lett. 88, 173901 (2002).

[39]. K. L. Corwin, N. R. Newbury, J. M. Dudley, S. Coen, S. A. Diddams, K. Weber, and R. S. Windeler, "Fundamental noise limitations to supercontinuum generation in microstructure fiber," Phys. Rev. Lett. 90, 113904 (2003).

[40]. J. M. Dudley, G. Genty, and S. Coen, "Supercontinuum generation in photonic crystal fiber," Rev. Mod. Phys. 78, 1135–1184 (2006).

[41]. X. Gu, L. Xu, M. Kimmel, E. Zeek, P. O'Shea, A. P. Shreenath, R. Trebino, and R. S. Windeler, "Frequency-resolved optical gating and single-shot spectral measurements reveal fine structure in microstructure-fiber continuum," Opt. Lett. 27, 1174–1176 (2002).

[42]. X. Gu, M. Kimmel, A. Shreenath, R. Trebino, J. Dudley, S. Coen, and R.Windeler, "Experimental studies of the coherence of microstructure-fiber supercontinuum," Opt. Express 11, 2697–2703 (2003).

[43]. Pedram Abdolghader, Adrian F. Pegoraro, Nicolas Y. Joly, Andrew Ridsdale, Rune Lausten, François Légaré, and Albert Stolow, "All normal dispersion nonlinear fibre supercontinuum source characterization and application in hyperspectral stimulated Raman scattering microscopy," Opt. Express 28, 35997-36008 (2020)

[44]. K. M. Hilligsøe, T. Andersen, H. Paulsen, C. Nielsen, K. Mølmer, S. Keiding, R. Kristiansen, K. Hansen, and J. Larsen, "Supercontinuum generation in a photonic crystal fibre with two zero dispersion wavelengths," Opt. Express 12, 1045– 1054 (2004).

[45]. E. R. Andresen, H. N. Paulsen, V. Birkedal, J. Thøgersen, and S. R. Keiding, "Broadband multiplex coherent anti-Stokesraman scattering microscopy employing photonic-crystal fibers," J. Opt. Soc. Am. B 22, 1934–1938 (2005).

[46]. M. Frosz, P. Falk, and O. Bang, "The role of the second zero-dispersion wavelength in generation of supercontinua and bright-bright soliton-pairs across the zero-dispersion wavelength," Opt. Express 13, 6181–6192 (2005).

[47]. M. Nakazawa, K. Tamura, H. Kubota, and E. Yoshida, "Coherence degradation in the process of supercontinuum generation in an optical fiber," Optical fibre Technology 4, 215 – 223 (1998).

[48]. K. Chow, Y. Takushima, C. Lin, C. Shu, and A. Bjarklev, "Flat super-continuum generation based on normal dispersion nonlinear photonic crystal fibre," Electronics Letters 42, 989–991 (2006).

[49]. K. P. Hansen, "Introduction to nonlinear photonic crystal fibers," Journal of Optical and fibre Communications Research 2, 226–254 (2005).

[50]. P. S. J. Russell, "Photonic-Crystal Fibers," J. Lightwave Technol. **24**, 4729-4749 (2006).

[51]. G. Agrawal, Nonlinear Fibre Optics (Academic Press, 2001).

[52]. E. A. Golovchenko, E. M. Dianov, A. M. Prokhorov, and V. N. Serkin, "Decay of optical solitons," JETP Lett. 42, 74–77 (1985).

[53]. Y. Kodama and A. Hasegawa, "Nonlinear pulse propagation in a monomode dielectric guide," IEEE J. Quantum Electron. 23, 510 – 524 (1987).

[54]. F. M. Mitschke and L. F. Mollenauer, "Discovery of the soliton self-frequency shift," Opt. Lett. 11, 659–661 (1986).

[55]. N. Akhmediev and M. Karlsson, "Cherenkov radiation emitted by solitons in optical fibers," Phys. Rev. A 51, 2602–2607 (1995).

[56]. X. Gu, M. Kimmel, A. Shreenath, R. Trebino, J. Dudley, S. Coen, and R. Windeler, "Experimental studies of the coherence of microstructure-fiber supercontinuum," Opt. Express 11, 2697–2703 (2003).

[57]. K. Mori, H. Takara, S. Kawanishi, M. Saruwatari, and T. Morioka, "Flatly broadened supercontinuum spectrum generated in a dispersion decreasing fibre with convex dispersion profile," Electron. Lett. 33, 1806–1808 (1997).

[58]. A. M. Heidt, "Pulse preserving flat-top supercontinuum generation in all-normal dispersion photonic crystal fibers," J. Opt. Soc. Am. B **27**(3), 550–559 (2010).

[59]. J. P. Gordon, "Theory of the soliton self-frequency shift," Opt. Lett. 11, 662–664 (1986).

[60]. P. S. J. Russell, ' Photonic-Crystal Fibers," J. Lightwave Technol. **24**, 4729-4749 (2006).

[61]. K. Saitoh, M. Koshiba, "Empirical relations for simple design of photonic crystal fibers" Opt. Express **13**, 267-274 (2005).

[62]. T. A. Pologruto, B. L. Sabatini, K. Svoboda, Biomed.Eng. Online 2,13 (2003).

[63]. M. Mohseni, C. Polzer, and T. Hellerer, "Resolution of spectral focusing in coherent Raman imaging," Opt. Express **26**(8), 10230– 10241 (2018).

[64]. A.M. Heidt, A. Hartung, H. Bartelt, The Supercontinuum Laser Source, (Springer, 2016).

Chapter 3

Introduction to Machine Learning

3. Introduction

Since ancient Greece, humans have always tried to build machines that could think [1]. From birth, humans obtain a lot of daily information by observation (hearing, seeing, etc.). All of this acquired knowledge is used as the logic behind every human decision made. To make a machine behave like a human, the same knowledge must be applied to computers. Implementing this task is one of the most challenging problems in computer science [1]. The term machine learning (ML) is referred to the task of extracting raw data patterns [1]. The main contributions of machine learning approaches to nonlinear optical microscopy are discussed in this chapter. To identify different materials in a heterogeneous sample based on their Raman spectra, we proposed use of a Convolutional Autoencoder (CAE) neural net. As we show, the trained CAE model was also able to denoise hyperspectral Stimulated Raman Scattering images obtained at low laser power input. Some basic machine learning concepts, such as the neural net, classification task, and convolutional neural net layer are briefly described in this chapter.

3.1 Definition of Machine Learning

Over the last decade, machine learning attracted much attention in fields for which there is a need for powerful analysis tools [1]. Machine learning techniques are made up of algorithms which allow a computer to learn from data; nevertheless, there is no universally accepted definition of machine learning, even among machine learning (ML) practitioners. In 1959 Arthur Samuel defined ML as a field of study which gives computers the ability to learn without being explicitly programmed [2]. He wrote a checkers playing program. What Arthur did was he programmed maybe tens of thousands of games against himself. The checkers playing programme then learned the correct and incorrect board positions over time by observing which board positions tended to lead to wins and which board positions likely to lead to losses. And ,finally, it learned to play checkers better than he. In 1998, Tom Mitchell came up with a new definition: a computer program is said to learn from experience E with respect to some tasks T and some performance measured P, if its performance of T, as measured by P, improves with experience E [3]. For instance, an email spam finder program can be trained due to removing spam email from the inbox. Here classifying emails as spam or not spam is called task (T); tracking your personal labelling of emails as spam or not is called experience (E), and performance of this program is (P). There are several types of machine learning algorithms, including supervised, semi-supervised, unsupervised, and reinforcement learning. We used supervised and unsupervised learning techniques on our datasets in this thesis.

3.2 Supervised Learning

Supervised learning is among the most common machine learning techniques. In supervised learning, the dataset used to train a model is called a labelled dataset. This means that, for every input dataset, we knew in advance what output the model was expected to produce. Basically, we feed the model with the dataset and by implementing of some optimization algorithm, the model tries to generate desirable output which is already known in supervised manner. As an example, we can train a model which could predict whether the input image is a cat or a dog: this is called binary classification. This can be done by giving a model thousands of images of dogs and cats with all images labelled accordingly. An objective function is required to train that model. A model also depends on some parameters which must be determined. These variable parameters, usually known as weights, are real numbers that can be viewed as 'knobs' which determine the model's input-output function [1]. The desired solutions, referred to as labels, are included in the training dataset provided to the algorithm. As mentioned above, in supervised learning the dataset is a collection of labelled examples $\{(x_i, y_i)\}_{i=1}^{N}$. Each element x_i among N (number of examples) is called a feature vector. A feature vector is a D-dimensional vector in which each dimension contains a value which describes the example data. That value is called a feature and is denoted as $x^{(j)}$, with $j = 1,...,D$, whereas y_i is called a label [1,4]. For instance, in a given dataset, if each example x in the dataset represent a person, then the first feature $x^{(1)}$ could contain the person's weight (mass), a second feature $x^{(2)}$ could contain the person's eye colour and so on. For all examples in the dataset, a feature at position j in the feature vector always contains the same type of information. This means that if $x_i^{(2)}$ contains a weight in some example x_i , then $x_k^{(2)}$ will also be a weight in every example x_k, where $k = 1,...,N$ in which N is number of examples. The label y_i can be either an element which belongs to a finite set of classes $\{1,2,...,C\}$, or a real number, or a more complex structure like a tensor. The goal of supervised learning is to utilize the dataset to create a model which takes a feature vector x as input and produces an output y that allows the label for this feature vector to be deduced. For example, a model built using a dataset of cancerous and healthy cell images may take a feature vector reflecting a cell's attributes as input and generate as output a probability of whether the cell is cancerous or not.

3.3 Unsupervised Learning

A dataset in unsupervised learning is a set of unlabeled examples $\{x_i\}_{i=1}^{N}$. Here x is a feature vector and the goal of machine learning is to learn useful properties from the structure of the dataset [1,4]. For example, in clustering, the model returns the cluster label for each feature vector in the dataset. One example where clustering is used is Google News. What Google News does is it daily looks at tens of thousands or hundreds of thousands of new stories on the web and groups them into cohesive news stories. There are a variety of unsupervised techniques available, including Principal Component Analysis [5], K mean Clustering [6], and Denoising autoencoders [7], which we shall cover in depth in this chapter. In this thesis, unsupervised models were used to classify different materials in the sample and improve image quality in our dataset (hyperspectral SRS images).

3.4 Supervised versus Unsupervised learning

In unsupervised learning, the model, by seeing many examples in a dataset, attempts to learn shared features amongst the examples: it tries to learn a feature probability distribution [1]. On the other hand, supervised learning is based on seeing lots of examples with associated values (labels) corresponds to that example: it learns how to predict a label from an example. The phrase "supervised learning" comes from a label given by an instructor who instructs the computer on what to perform [1]. Unsupervised learning, on the other hand, has no instructor. The distinction between supervised and unsupervised is frequently blurred [1].

3.5 Dataset

A dataset is a collection of examples, each of which is made up of features. A design matrix or model matrix is a commonly used method of representing a dataset. In each row of a design matrix, there is a different example. A separate feature is assigned to each column of the design matrix [1]. For instance, consider a dataset of 150 examples with each having four features. This can be represented by a design matrix $X \in R^{150 \times 4}$.

3.6 Machine learning models

Using a machine learning technique necessitates the creation of a model which has been trained on a dataset and can subsequently process more data to produce predictions. These algorithms can help a computer program perform better at specific tasks by showing it examples. Machine learning models come in a variety of designs, each with its own set of applications. Below, some of the most common machine learning models will be introduced: we applied these models in this thesis.

3.6.1 Linear Regression

In linear regression the goal is to build a model which can take an input $x \in R^n$ which is introduced as a feature vector and then predict an output $y \in R$. The dataset is a collection of examples $\{(x_i, y_i)\}_{i=1}^{N}$, where N is the number of examples in the dataset and x is a n-dimension vector (feature vector). The output of linear regression is a linear function of the input. The output can be defined as:

$$\hat{y} = w^T x .$$

$$(4.1)$$

where, $w \in R^n$ is the vector of parameters and T means transpose operator.

Parameters are values which, like a knob, determine how the model behaves. Before summing up all of the features' contributions, the feature vector and vector of parameters are multiplied element-by-element. We can consider these products as weights which determine the impact of each feature on the prediction. In this model, the task is predicting y from x by outputting $\hat{y} = w^T x$. Assume we have a design matrix with "m" example inputs which will not be used for training but only be used to evaluate how well the model works. In addition, we have a vector target which provides the proper value of y for each example. We call this dataset the test dataset because it will only be used for evaluation of the model. We refer to the design matrix of inputs as $x^{(test)}$ and the vector of regression targets as $y^{(test)}$ [1].

We have an input dataset with some examples, and an output dataset with the input and some model parameters which we want to identify ($weights$) in order to get the best predictions. Calculating the model's mean square error (MSE) on the test dataset is one measure of its performance. If $\hat{y}^{(test)}$ gives the prediction of the model on the test dataset, then the MSE is given by [1]:

$$MSE_{test} = \frac{1}{m} \sum_{i=1}^{m} (\hat{y}^{test} - y^{test})_i{}^2 . \tag{4.2}$$

In order to train a machine learning model, an algorithm must be designed which adjusts the weights "w" in order to minimize the MSE_{test} when the algorithm observes a training set $\left(x^{(train)}, y^{(train)}\right)$ during its training step [1]. The MSE is defined as an objective or cost function. The purpose of training a model in machine learning is to minimize the cost function [1]. To minimize the cost function, we can solve for its zero gradient:

$$\nabla_w MSE_{train} = 0 . \tag{4.3}$$

$$\nabla_w \frac{1}{m} \left\| \hat{y}^{(train)} - y^{(train)} \right\|^2 = 0 . \tag{4.4}$$

$$\nabla_w \left(x^{(train)}w - y^{(train)}\right)^T \left(x^{(train)}w - y^{(train)}\right) = 0. \tag{4.5}$$

After taking the gradients, we will have:

$$w = \left(x^{(train)T} x^{(train)}\right)^{-1} x^{(train)T} y^{(train)} . \tag{4.6}$$

Equation (4.6) is called the Normal Equation. There is an additional parameter termed "bias" in linear regression and other machine learning methods. It appears as an intercept in linear regression. In the modified version of the linear regression model looks like this:

$$\hat{y} = w^T x + b. \tag{4.7}$$

The bias parameter has no effect on the mapping function's overall form. Adding a bias term keeps the model linear, as demonstrated in equation (4.7), The mapping from features to prediction, on the other hand, is now an affine function. The term bias refers to the fact that in the absence of any input, the model's output is biased to be "b" [1].

3.6.2 Cost function

The performance of machine learning models is evaluated using the cost function/objective function. It indicates how well the model predicts or makes decisions for a given set of parameters. Many machine learning techniques use optimization to determine the model's parameters by minimizing the cost function. For instance, in the linear regression (4.6.1), the cost function is defined as the mean square error, and by solving the Normal equation (4.6), we could determine the weights $'w'$ of the model. Different cost functions exist depending on the application for which the model will be trained [1]. The MSE cost function is typically employed in regression models, whether linear or nonlinear. However, another typical cost function, cross-entropy, is employed for classification problems. As a result, choosing which cost function to utilize is largely dependent on the model application. There are various methods for optimization, the most common of which being gradient descent algorithms. This chapter will describe how 'gradient descent' and 'stochastic gradient descent' work; the next section will cover a more advanced version called 'mini-batch gradient descent.'

3.6.3 Gradient descent optimization

Gradient descent is one of the most widely used methods for machine learning model optimization. It's an iterative optimization approach for determining the weights value which minimize the cost function. Suppose we have a function $y = f(x)$, where both y and x are real numbers. The first derivative of this function is expressed as $f'(x)$. The derivative of the function gives the slope of that function at the point x. It explains how to scale a tiny change in input produces a change in output [1]:

$$f(x + \epsilon) \approx f(x) + \epsilon f'(x). \tag{4.8}$$

As a result, the derivative is valuable for minimizing functions since it shows us how to adjust x to alter y by a small amount. It is known that $f(x - \epsilon * sing(f'(x)))$ is always less than $f(x)$ for small value of ϵ. Thus, we can reduce $f(x)$ by moving x in small steps with opposite sign of the derivative. This technique is called gradient descent. See Fig. 3.1 for an example of this technique [1].

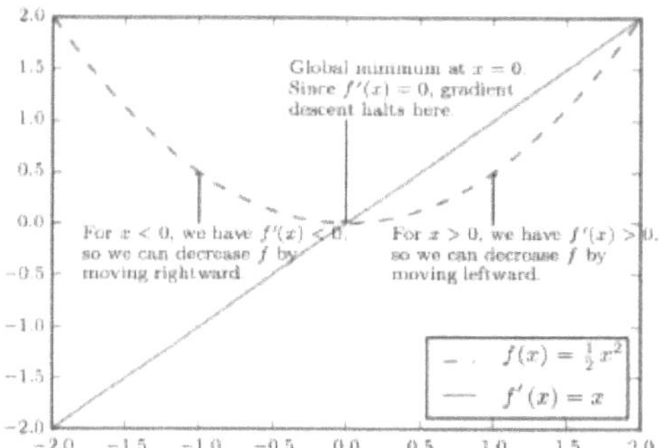

Fig. 3.1: Gradient descent. An illustration of how gradient descent algorithm uses a function's derivatives to follow the function downhill to a minimum. The figure is peaked from reference [1].

When $f'(x) = 0$, the derivative gives no information about which direction to move. Points where, $f'(x) = 0$ are known as critical points. Three types of critical points exist. A local minimum: a point where $f(x)$ is lower than all adjacent points, making it impossible to decrease $f(x)$ by making small steps. A local maximum: a point where $f(x)$ is higher than all adjacent points, so it is impossible to increase $f(x)$ by making small steps. Finally, saddle points, which are neither maxima nor minima [1], they are all shown in Fig 3.2.

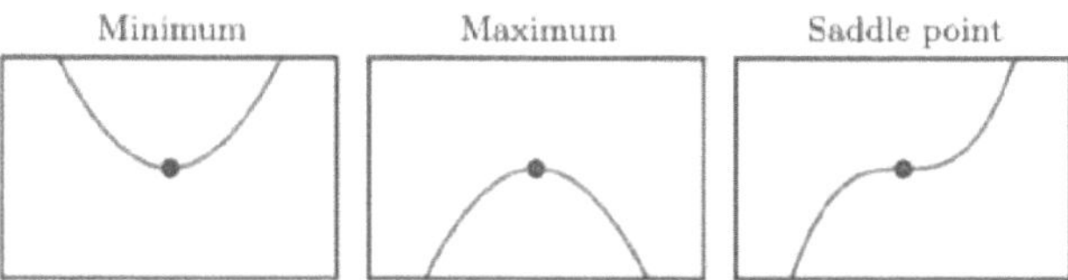

Fig. 3.2: Three types of critical point in one dimension. A critical point is a point that ist' derivative equals to zero. Such a point can be a local minimum, local maxima or a saddle point, figure is peaked from reference [1].

The global minimum refers to the point where the function's absolute lowest value occurs. A single global minimum or numerous global minima are both possible. Local minima that are not globally optimal are also possible, are not ideal, and many saddle points are surrounded by relatively flat regions [1]. All of these factors make optimization challenging, especially when the function's input is multidimensional, see Fig. 3.3 [1].

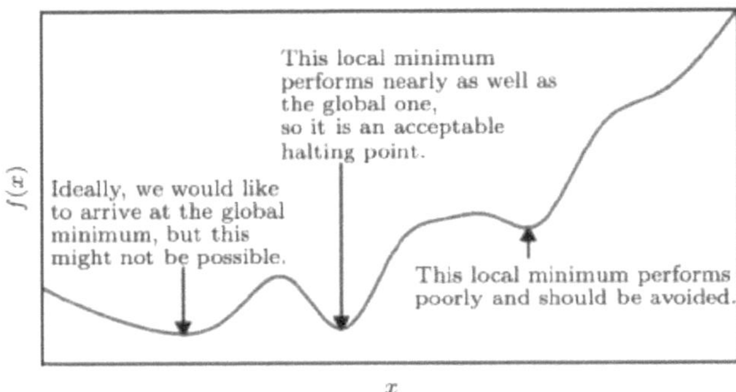

Fig. 3.3. Representation of some local minimums and a global minimum. Optimization algorithms may fail to find a minimum when there are several local minimums, figure is peaked from reference [1].

We must employ the partial derivative idea in machine learning, since the model is practically dependent on several parameters. The partial derivative $\frac{\partial f(x)}{\partial x_i}$ measures how f changes if only the variable x_i changes at point x. In a model with multiple variables, gradient is applied where the derivative is with respect to a vector. The gradient of f is the vector containing all the partial derivatives, denoted $\nabla_x f(x)$ [1].

The directional derivative in direction u (a unit vector) is the slope of the function in direction u. The directional derivative is the derivative of the function $f(x + \alpha u)$ with respect to α, evaluated at $\alpha = 0$. To minimize f we would like to find a direction that f decreases. This can be done by [1]:

$$min_{u,u^T} \; u^T \nabla_x f(x). \qquad \text{(T means transpose operator)} \qquad (4.9)$$

$$min_{u,u^T} \; ||u||_2 ||\nabla_x f(x)||_2 \cos(\theta). \qquad (4.10)$$

Where θ is angle between the unit vector of u and the gradient. Since $|u| = 1$, by ignoring factors that are not related to u, the equation 4.10 simplifies to: $min_u \cos(\theta)$. This minimizes when u points in the direction opposite to the gradient. Thus, f can be decreased by moving in the direction of the negative gradient [1]. This is known as the method of steepest descent or gradient descent algorithm. Thus, we should substitute the x feature with a new value [1]:

$$x := x - \varepsilon \nabla_x f(x). \tag{4.11}$$

where ϵ, is the learning rate which is a positive scalar defining the size of each step-in minimization process [1]. The learning rate could be determined empirically, and sometimes we can solve the step size that makes the directional derivative zero [1]. Gradient descent converges when all gradient elements approach to zero, or quite close to zero [1].

3.6.4   Gradient descent variants

Gradient descent optimization can be implemented in three different ways depending on how the input dataset is fed into the model: 1- Batch gradient descent, 2- Stochastic gradient descent and 3- Mini batch gradient descent. In batch gradient descent (BGD), all the training data is taken into consideration to take a single step. Batch gradient descent computes the gradient of the cost function with respect to the entire training dataset's model parameters. Batch gradient descent will always converge to the global minimum for convex error surfaces (no local minimum exists) and a local minimum for non-convex surfaces [1]. This method has a significant disadvantage in that it is too slow to implement. Datasets frequently have a large number of examples with many features. As a result, a large dataset might make even a single iteration very long to compute. To tackle long-time computing optimization, optimizing the cost function by batch gradient descent, stochastic gradient descent (SGD) can help. Here, we consider one sample at a time, in order to take a single step. Because we only consider one sample at a time in SGD, the cost function will fluctuate throughout the training process and will not necessarily decrease unless the model is trained for an extended amount of time [1]. BGD should be utilized for cost functions with a lower local minimum in general. When the dataset is large, though, SGD can be used. Generally, SGD converges faster than BGD. One of the drawbacks of SGD is that we cannot implement the vectorization method because it just uses one example at a time. The vectorization method is a computer approach which uses rapid matrix multiplication techniques to make calculations extremely fast. A combination of SGD and BGD is required to overcome the limitations of both methods [1]. We can split the dataset into small groups called mini-batches (subsets) instead of using the complete dataset at once or only utilizing one sample at a time. Doing this helps to achieve the advantages of both former variants methods. In Table 3.1, the steps required to train the model in one iteration are summarized. Steps 1-4 then repeat over some number of iterations, called the epoch, until the training error converges. Convergence of the training error can be observed in a learning curve like Fig. 3.6. Note that gradient descent and its variants are not machine learning algorithms. They are solvers of minimization problems in which the function to minimize has a gradient.

Batch gradient descent	Stochastic gradient descent	Mini-batch gradient descent
1-Take the entire dataset.	1-Take one example from the dataset.	1-Pick a mini-batch
2-Feed it to the model.	2-Feed it to the model.	2-Feed it to the model.
3-Calculate its mean gradient.	3-Calculate its gradient.	3-Calculate its mean gradient.
4-Use the calculated mean gradient in step 3 to update the model parameters.	4-Use the calculated gradient in step 3 to update the model parameters.	4-Use the calculated mean gradient in step 3 to update the model parameters.
5-Repeat steps 1-4 for all the examples in the dataset.	5-Repeat steps 1-4 for all examples in training dataset.	5-Repeat steps 1-4 for the created mini-batches.

Table 3.1. Summary of all steps that are required to take in one iteration to train a model.

3.7 A practical example of training a model with gradient descent

To give an instructional example of training a simple model (linear regression), a synthetic dataset with 200 examples is generated. The dataset wherein each (x_i, y_i) is considered as a sample is shown in Fig. 3.4. The model was then trained to find the best linear fit to this dataset. The cost function is the MSE and the gradient descent algorithm (batch gradient descent) is performed to minimize the cost.

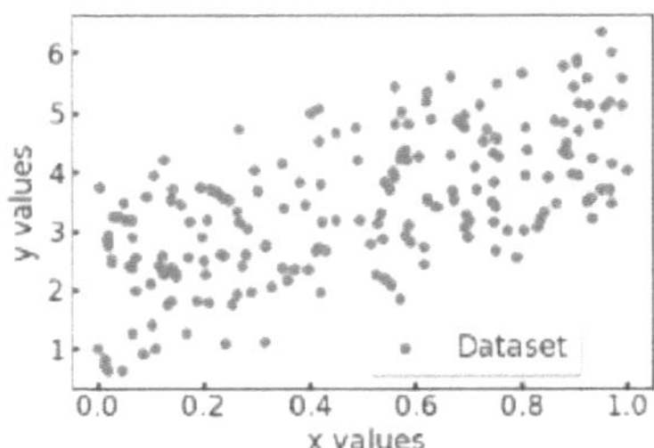

Fig. 3.4. An arbitrary synthetic dataset consists of 200 examples. The goal of training a linear regression model is to find the best line to fits the samples.

The linear regression model looks like $f(x) = wx + b$. We don't know the optimal values of b or w are and want to determine these from the dataset after training the model. To do this, we look for values for w and b which minimize the cost function, defined below:

$$cost = l = \frac{1}{N}\sum_{i=1}^{200}(y_i - (wx_i + b))^2. \tag{4.12}$$

To apply the gradient descent algorithm to minimization of the cost function, it is required to calculate the first derivative of the cost function with respect to its parameters (here w and b).

$$\frac{\partial l}{\partial w} = \frac{1}{N}\sum_{i=1}^{200} -2x_i(y_i - (wx_i + b)). \tag{4.13}$$

$$\frac{\partial l}{\partial b} = \frac{1}{N}\sum_{i=1}^{200} -2(y_i - (wx_i + b)). \tag{4.14}$$

Gradient descent proceeds in epochs. An epoch consists of using the training set entirely to upgrade each parameter (BGD). It is also required to initialize the models' parameters. Here zero value is given to our linear regression model parameters (w and b). However, for complex models in which thousands of parameters are involved, the parameters' initialization may remarkably affect the training; this is why there are different techniques for initializing the model parameters. At each epoch, the training process updates w and b by using partial derivatives (4.13-14). The learning rate (α) controls the step size at each update:

$$w := w - \alpha\frac{\partial l}{\partial w}. \tag{4.15}$$

$$b := w - \alpha\frac{\partial l}{\partial b}. \tag{4.16}$$

This process of updating parameters repeats until the cost function converges. Typically, this iterative process requires many epochs until the values of w and b don't change much after each epoch. In Fig 3.5 we show the training process at some epochs. At the higher epochs, the linear curve that is fit to the data doesn't change much. This model is built on Python language and the source codes are provided in appendix 1. The training time was less than one minute.

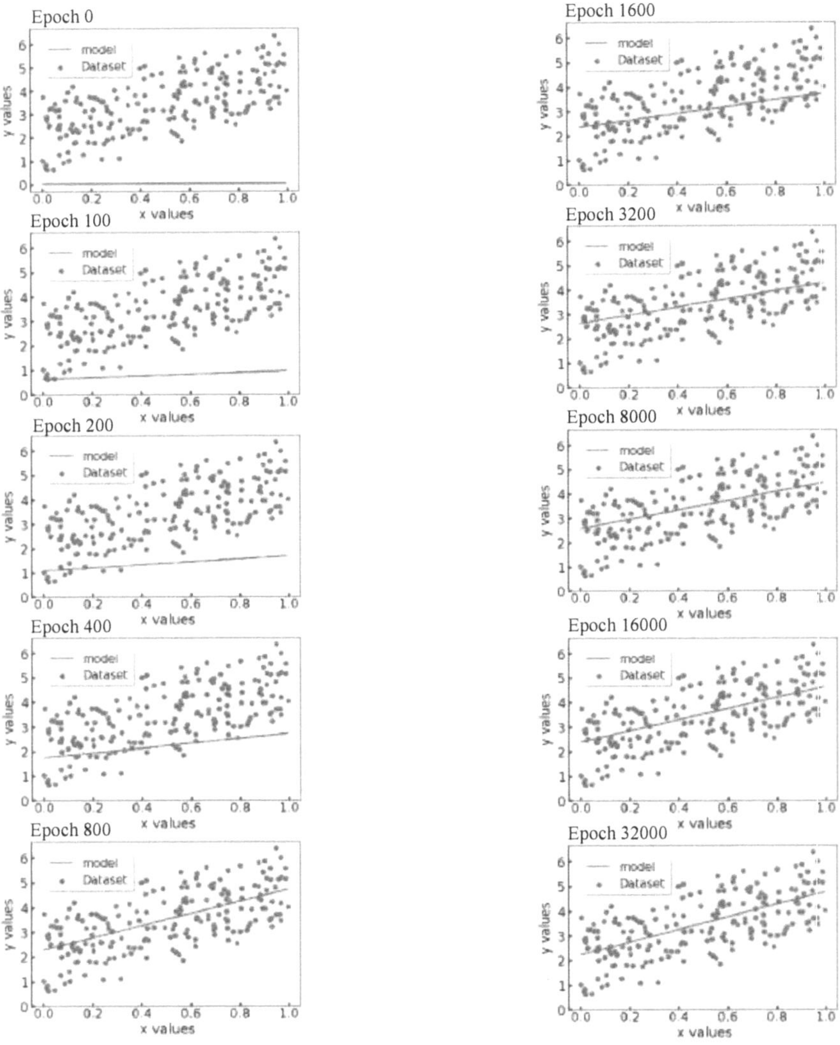

Fig. 3.5. The evolution of the regression line fitting on the samples during the training at some epoch.

Finally, after 32000 iterations the model predicts that w and b should have the values of 2.55 and 2.22, respectively. The learning curve which demonstrated loss (cost function values) versus the number of iterations is shown in Fig. 3.6. After a specific epoch number, the model doesn't learn more, which means that continuing the training time won't give us a better model. After approximately 10^3 epochs, the model doesn't learn anymore and even might lead to an overfitting problem, discussed in the next sections.

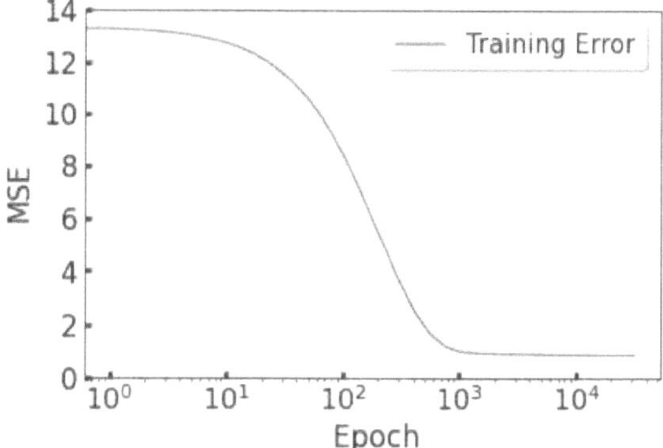

Fig. 3.6. Learning curve: Training error (loss) versus epoch number.

3.8 Generalization

The crucial challenge in machine learning is that the trained model must not only work on the training dataset. The model also has to work on a test dataset. A test dataset is a dataset that has not been seen by the model during the training [1]. Generalization is the name which is given to this capacity. The error (cost function) is calculated using the training dataset as input. So far, we've only discussed an optimization problem. Machine learning, on the other hand, is not the same as optimization. The distinction is related to a generalization error, which means that the trained model must also work on a test dataset [1]. The error of the test dataset has to be as low as the error of the training dataset. Usually, the test dataset was collected separately from the training dataset. An important question that one might ask is why a trained model can work on a test dataset that the model has not seen before. The field of *statistical learning theory* provides some answers [1,8-10]. Statistical learning theory discusses how to find a predictive function from a set of data [1].

Arbitrary collections of training and test datasets aren't helpful. However, by making some assumptions about collecting the training and test dataset, we can improve the test dataset's error by adjusting the model's parameters trained with the training dataset [1]. A probability distribution over datasets generates the training and test dataset called the data generating process. A set of assumptions are made, known as the i.i.d. assumptions [1]. The examples in each dataset are assumed to be independent of one another. The training and test sets are also distributed in the same way, using the same probability distribution. This assumption allows us to use a probability distribution to explain the data generation process over a single example [1]. Every training and test example is then generated using the same distribution. That underlying shared distribution is called the data-generating distribution, denoted by p_{data} [1]. The expected training error of a randomly selected model is equal to the expected test error of that model, according to the relationship that can be detected between training error and test error. This result is based on the assumption we made: both the test and training datasets are sampled from a same probability distribution [1]. However, we must keep in mind that when using a machine learning algorithm, we do not define the parameters in advance and then sample for both datasets (train and test) from the probability distribution. To reduce the training error, we sample the training set and use it to train the model by modifying the model's parameters. The test set is then sampled and fed into the model. The expected test error in this process is greater than or equal to the expected training error. Two factors are involved in making a machine learning model: 1- the training error has to become as small as possible, and 2- the gap between the test error and training error must be small [1].

3.9 Overfitting and underfitting

The two aspects discussed in the preceding section correspond to machine learning's two key obstacles. Overfitting and underfitting are terms used to describe these situations. When the difference between the training error and the test error is too large, overfitting occurs (perfect prediction on the training dataset and poor prediction on the test dataset). Underfitting, on the other hand, occurs when the model is unable to achieve a low enough error result on the training dataset.

Bias and high variance are terms used to describe underfitting and overfitting, respectively [1]. There are some strategies for dealing with the issues of underfitting and overfitting [1]. In the case of underfitting, we can try a more complex model or engineer input features. We can try a simpler model, reduce the dimensionality of the dataset's examples (for example, using the principal component analysis methodology), add more training data, or apply regularisation techniques to avoid overfitting. Fig. 3.7. demonstrates three models that were trained to fit a curve on a dataset. Underfitting, best fit and overfitting are shown [1].

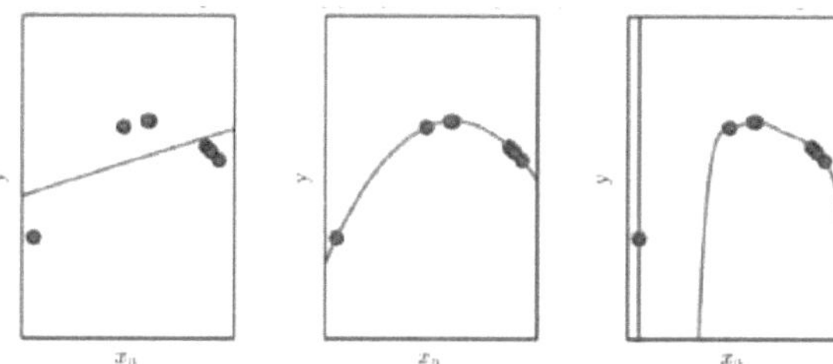

Fig. 3.7. Synthetic data was used to create the training/test dataset. Left fit is a linear function that is underfitting the dataset. The best fit is the centre fit, which is a quadratic function. The right side fit is a polynomial function that has been overfitted in the dataset. This figure is adapted from reference [1].

Adjusting the model's capacity is one method of avoiding the problem of overfitting or underfitting. The capacity of a model is defined as its ability to perform a wide range of tasks [1]. Models with low capacity have a difficult time fitting the training data. Models with high capacity, on the other hand, can be overfitted by memorizing properties from the training dataset [1]. The hypothesis space is a set of solutions which could be selected by a model as a function for mapping input features to the output of a machine learning model. We can adjust a model by changing its capacity. For example, the hypothesis space of linear regression is made up of all linear functions of the input. We may generalize the linear regression hypothesis space by include polynomials, which boosts the model's capacity [1].

A machine learning algorithm's ideal performance could be achieved when an appropriate capacity is chosen based on the complexity of the problem and the given dataset to the model [1]. Models with inadequate capacity cannot handle difficult tasks; however, models with a high capacity can handle complicated tasks at the penalty of overfitting when the model's capacity exceeds its requirements. Fig. 3.8 represents error versus capacity of a machine learning model for training error and generalization error [1]. Plotting a graph representing both training and test error versus model capacity can be used to determine a model's optimum capacity.

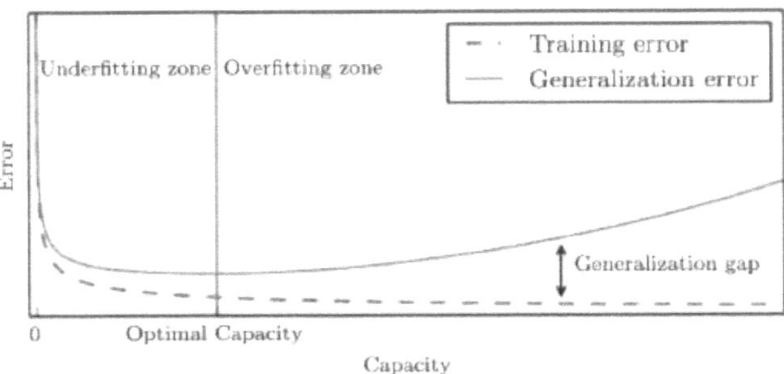

Fig. 3.8. Capacity and error have a typical connection. Both training and test error are high when the model capacity is low (underfitting zone). The training error lowers as model capacity rises; however, the gap between the training and generalization errors grows, causing the model to enter an overfitting zone, figure is peaked from reference [1].

3.9.1   Regularization

Regularization is a phrase which refers to techniques that force the learning process to build a simpler model. Regularization often results in a tiny increase in bias but dramatically reduces variance [1,4]. The bias-variance trade-off is a term used in the literature to describe this problem [1,4]. The objective function is modified to create a regularization model by adding a penalizing term whose value is higher when the model is more complicated [1,4]. For instance, in linear regression, a modified objective function would be like this:

$$min_{w,b}[C\,|w| + \frac{1}{N}\sum_{i=1}^{N}(y_i - (wx_i + b))^2].$$

(4.17)

where $|w| = \sum_{j}^{D}|w^{(j)}|$, and C is a hyperparameter that controls the influence of the regularization term. By setting $C = 0$, the model becomes a non-regularized linear regression (4.12). On the other hand, when we set C to a large value then the training procedure tries to set most of the weights $w^{(j)}$ to a small value or even zero in order to keep the objective function small (the purpose of training a model) [4], which leads to the underfitting problem. Value of C has to be set in a way that doesn't much increase the bias but reduces the variance to a level reasonable, allowing the problem to be solved. The described example (4.17) used a regularization method which is called L1 regularization. Another regularization is named L2 is shown in 4.18:

$$min_{w,b}[C\,||w||^2 + \frac{1}{N}\sum_{i=1}^{N}(y_i - (wx_i + b))^2].$$

(4.18)

where $||w|| = \sum_{j}^{D}(w^{j})^{2}$. (The feature vector x is assumed to be a D dimensional vector)

3.10 Feature engineering

To train a model to do a task, we must first create a dataset. Feature engineering is the process of converting raw data into a dataset. In most cases, feature engineering is a time-consuming procedure which necessitates a significant amount of effort on the part of the data analyst. For example, when using a neural net to classify images, it is necessary to transform the range of pixel values to a specified range, such as (0-1), as well some other pre-image processing. All of these processes are performed in order to train a model more quickly and with better outcomes. One practical example in this thesis is in the training of an autoencoder model. The error in the training and test datasets was extremely high; however, the error was dramatically reduced simply by normalizing the input dataset and introducing a pre-processing step. Below are some important examples of feature engineering [4].

3.10.1 Normalization

Converting the actual range of input values into a standard range of values, typically in the [0-1] or [-1,1] range is called normalization. Normalization can lead to a boost in learning speed [4]. Furthermore, it is beneficial to make sure that our inputs are approximately in the same small range to prevent the common issue that computers have when working with extremely small or extremely large numbers [4]. Normalization can be done as:

$$\hat{x}^{(j)} = \frac{x^{(j)} - min(x^{(j)})}{max(x^{(j)}) - min(x^{(j)})}.$$

$$(4.17)$$

3.10.2 Standardization

The process of rescaling feature values is known as standardization (z-score normalization). They have the characteristics of a standard normal distribution, with a mean of 0 (averaged across all examples in the dataset) and a standard deviation of 1 from the mean. Standardization of features is calculated as follows:

$$\hat{x}^{(j)} = \frac{x^{(j)} - mean(x^{(j)})}{max(x^{(j)}) - min(x^{(j)})}$$

$$(4.18)$$

Most learning algorithm usually benefits from feature rescaling. There exist many libraries which automatically take care of the feature engineering process, depending on the task given to the model. Normalization and standardizations in neural net models depend significantly on the activation function of the model.

3.11 k-means clustering algorithm

Data clustering is a data analysis approach which groups objects that have similar features. A common unsupervised clustering algorithm is the k-means clustering algorithm [4]. The following explaions how the k-means algorithm works. The number of clusters (k), known as centroids, must first be determined. The cluster's centre is called a centroid which could be a real or imaginary number. Finding the right number of clusters (centroids) for a dataset is usually a case of trial and error. Some mathematical procedures, such as the elbow technique, can also be used to determine the number of centroids. The k-means algorithm is implemented by starting with the first group of randomly selected centroids, which serve as the starting points for each cluster. The distance between each dataset example and each centroid is then calculated using a measure such as the Euclidean distance. The nearest centroids are then allocated to each example. The average distance of the examples in the cluster with regard to the centroid is then calculated for each centroid, and the calculated average becomes the centroid's new location. The distance between examples and the calculated average (new centroid) is calculated once more, and each example is assigned to the centroid with the shortest distance. This iterative procedure continues until the algorithm converges, which means the location of centroids remains fixed. Briefly, the main element of the k-means algorithm works by a two-step process called expectation-maximization. The expectation step assigns each data point to its nearest centroid. The maximization step computes the mean of all the points for each cluster and sets the new centroid, as explained below:

1- Specify the number k of clusters to assign.
2- Randomly initialized k centroids.
3- **Repeat:**
      expectation: Assign each point to its closest centroid.
      maximization: Compute the new centroid (mean) of each cluster.

   Until the centroid positions do not change.

To give an instructional example of how the k-means algorithm works, a synthetic dataset with 20 examples was generated. The dataset had two features called X1 and X1, which are shown in Fig. 3.9. The expectation-maximization process over two iterations presented in Fig. 3.10. After two iterations, the location of centroids didn't change, implying the k-means algorithm converged.

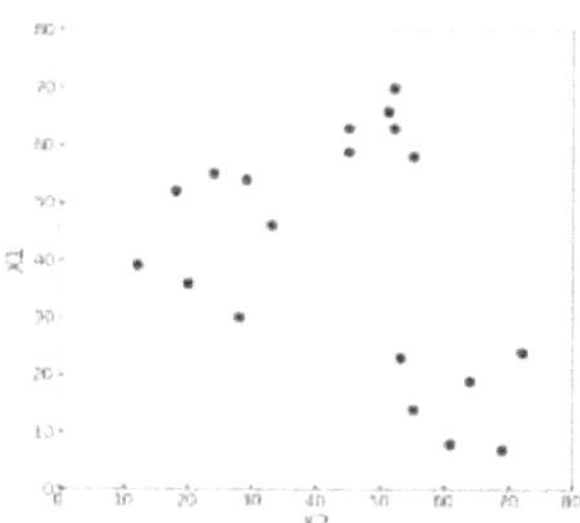

Fig. 3.9. A synthetic dataset with 20 examples. Each example depends on two features X1 and X2.

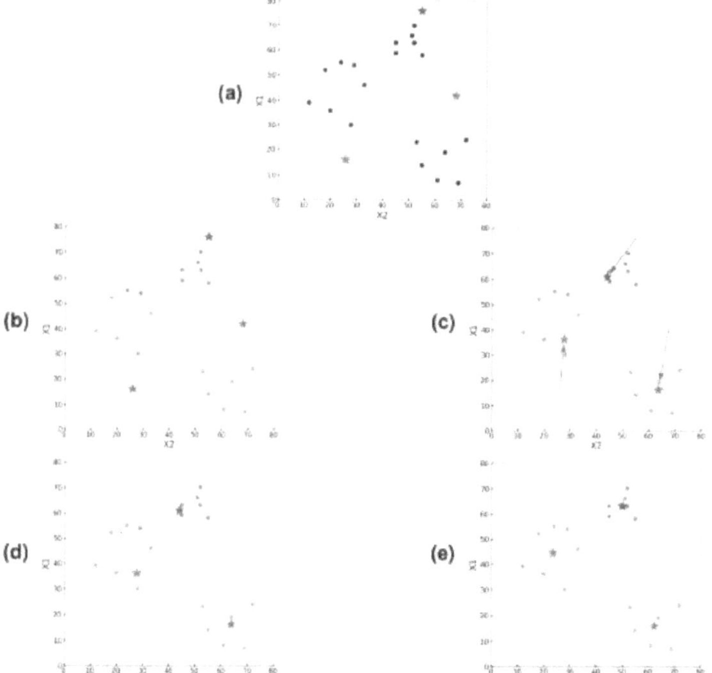

Fig. 3.10. Demonstration of how the k-means algorithm works. (a): initialization step, Randomly initialization of k (k = 3 here) centroids in feature space. Centroids indicate by colored stars. (b): expectation step, assigning each point to its closest centroid. (c): maximization step, computing the new centroids (mean) of each cluster and replacing the new centroids to the calculated ones. (d): repeat the expectation step. (e): repeat the maximization step.

3.12 Principal Component Analysis (PCA)

Large datasets are becoming more frequent in many fields and can be challenging to interpret. PCA is a statistical technique that finds the dependencies between the dataset variables and then reduces the dataset variables dimension to a few that contain most information.

3.12.1 Implementation of PCA algorithm

1- Normalization of the dataset.
2- Covariance matrix computation.
3- Compute the eigenvectors and eigenvalues of the covariance matrix to identify the principal components.
4- Building feature vector.
5- Recast the data along the principal component axes.

The first step equalizes the contribution of each variable in the dataset by normalizing the range of variables values. Standardization is necessary before PCA because data analysis is very sensitive to the variances of the initial variables. You can imagine a variable with a larger variance range would dominate in comparison to variables with a small range of variance (for instance, a variable within a variance range of 0 to 1000 will dominate over a variable with a variance range of 0 to10). As a result, converting the data to comparable scales can help avoid this situation. To do so, simply subtract the mean from each variable's value and divide by the standard deviation.

Because standard deviation and variance only work on one dimension, you could only compute the standard deviation for each dimension of the data set separately from the others. However, having a similar measure to evaluate how much the dimensions differ from the mean in relation to each other is useful. Computing the covariance is such a measurement. It illustrates whether or not there is a link between distinct dimensions of a dataset. The Covariance between two variables X and Y is defined in equation 4.19.

$$Cov\,(X,Y) = \frac{\sum_{i=1}^{n}(X_i - \bar{X})(Y_i - \bar{Y})}{n-1}. \tag{4.19}$$

It's important to note that covariance is always calculated between two dimensions. If we have a dataset with more than two dimensions, we can calculate more than one covariance. Calculating all possible covariance values between all of the different dimensions and putting them in a matrix called a covariance matrix is an useful approach. Here is an example. We'll make up the covariance matrix for an imaginary 3-dimensional dataset, using the usual dimensions x, y and z. Here, the covariance matrix has 3 rows and 3 columns as shown in 4.2.

$$C = \begin{pmatrix} cov\ (x,x) & cov\ (x,y) & cov\ (x,z) \\ cov\ (y,x) & cov\ (y,y) & cov\ (y,z) \\ cov\ (z,x) & cov\ (z,y) & cov\ (z,z) \end{pmatrix} \qquad (4.2)$$

In order to find the principal components (new variables to represent the dataset), it is required to calculate the eigenvectors and eigenvalues of the covariance matrix. New variables (principal components) are generated as linear combinations of the original variables. The new variables are uncorrelated as a result of these combinations. The first (largest eigenvalue) principal components contain the most information about the dataset in comparison to the later (smaller eigenvalue) principal components. If we sort the principal components based on their importance, the first ones are most useful in representing the dataset. It is worth mentioning that the principal components don't have any real meaning because they are just constructed from linear combinations of the original variables of the dataset. However, principal components show the maximum direction of the variance of a dataset. The principal components in order of significance can be established by ranking the calculated eigenvectors in order of their eigenvalues, from highest to lowest. After calculating the eigenvalues and eigenvectors, we decide whether to preserve all of these components or to discard those with low eigenvalues and build a matrix of vectors called a *Feature vector* with the remaining ones. So, the feature vector is simply a matrix whose columns are the eigenvectors of the kept components. Finally, the dataset is reoriented from the original axes to the principal component representation. This is accomplished by multiplying the original dataset's transpose by the feature vector's transpose.

3.13 Neural network

The neural network is an interesting branch of machine learning. It was inspired by the complex functionality of human brains which process information in parallel using on the order of one hundred billion linked neurons [1-4,12]. A neural network consists of an input layer of neurons (nodes, units) which input datasets fed to them, some hidden layers of neurons and a final layer of output neurons. Fig. 3.11 represents a typical architecture where weights are connecting different neurons. The weight (lines connecting neurons) is a numerical value assigned to each connection. Upon feeding the dataset to the neural net, if the output of any particular node exceeds a certain threshold value, that node becomes active and transfers data to the network's next layer. In the below threshold case, no data is sent to the next layer of the network [11]. The threshold is set by the activation function, discussed below.

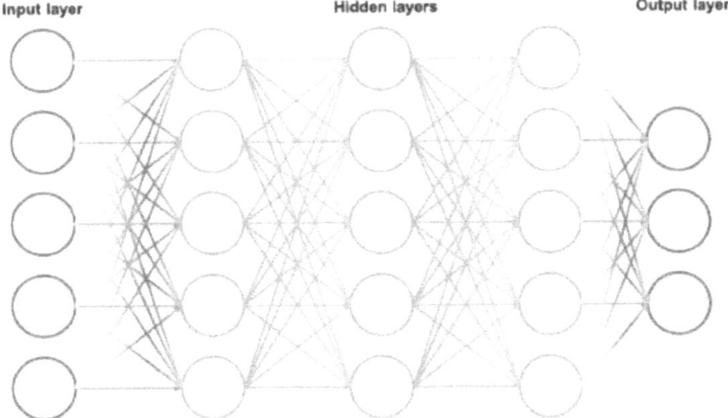

Fig. 3.11. The architecture of a typical neural net with three hidden layers (green circles). Neurons (nodes) represent by circles, and arrows that connect neurons represent weights. Neural nets rely on training data to learn and improve their accuracy over time to do a given task.

The output, h_i, of neuron i in the hidden layer is [12]:

$$h_i = \sigma\left(\sum_{j=1}^{N} W_{ij}x_j + b_i\right). \tag{4.19}$$

Where, $\sigma()$ is called activation function, N is the number of input neurons (in Fig. 3.11, N = 5 for the first four layers), W_{ij} the weights, x_j inputs to the neurons, and b_i the threshold (bias) terms of the hidden neurons. The activation function's aim is to bind the neuron's value so that divergent neurons do not paralyze the neural net and to introducing nonlinearity into the neural net [12]. Weights in neural networks, as in other machine learning models, are thought of as knobs which control the neural net's output (model). In the training of a neural net, all of the inputs are multiplied by their respective weights and then added together. The output is then sent through an activation function to determine the output. If the output surpasses a certain value (threshold), the node is activated and data is passed to the network's next layer. This results in the output of one node becoming in the input of the next node. This neural net is known as a feedforward network since data is passed from one layer to the next [12-13]. Before the training begins, the activation function is specified, and it is commonly a nonlinear function depending on the feature values in the input dataset. A common example of the activation function is the sigmoid function defined as:

$$\sigma(x) = \frac{1}{1+e^{-x}}. \tag{4.20}$$

Arc tangent and hyperbolic tangent are two other possible activation functions. They have a similar response to the inputs as the sigmoid, but their output range is different. Other types of activation functions, such as ReLu or Leaky ReLu, can speed up the training process. However, the choice of activation function depends on the input dataset. Similar to all machine learning models, a neural network is trained by a set of input data called a training set. The desired outputs of the training data are known in a supervised method; the training seeks to minimize errors by altering the weights between the connected neurons. The backpropagation algorithm is commonly used to adjust the values of the weights in neural networks. Backpropagation is a gradient-based technique which updates the weights of a neural network using the derivative chain rule. The backpropagation technique was first established in the 1970s, but its significance was not fully realized until it was used to train a neural network [14].

3.13.1 Types of Neural networks

Neural networks can be classified into different types which are used for various purposes. The most common types are feedforward neural nets which were discussed above, convolutional neural networks (CNNs), recurrent neural networks (RNNs) and long short-term memory (LSTM). In this thesis, we will discuss CNNs in more detail.

3.13.2 Convolutional neural networks

Convolutional neural networks are a type of neural network comparable to feedforward nets but have fewer parameters than artificial neural networks (ANNs). The computational complexity required to compute data is one of the major drawbacks of ANNs. In order to demonstrate why CNNs are advantageous for speeding up calculation time, let's assume that a network receives 32x32 raw RGB (red, green, blue) pixels as input [15]. As a result, 32*32*3 weights are required to connect the input layer, which is a 32*32 pixels RGB (3 channels) image, to one neuron. Adding one more neuron into the hidden layer increases the number of weights to 32*32*3*2, which means ~ 6000 weight parameters are required to connect the input to just two neurons. Two neurons may be insufficient for image processing tasks such as classification. We can connect the input image to the next layer's neurons with the same height and width numbers to make it more efficient. This network needs 32*32*3 by 32*32 weight connections, meaning that around 3 million parameters are required for training.

Rather than a full connection between all input layer neurons and the next hidden layer, it is good to look at local regions in the image (data). In this case, the hidden neurons in the following layer only receive input from a small section of the previous layer. For example, a neuron could be connected to only 5*5 neurons from the previous layer, rather than all of them. Thus, if we require 32*32 neurons in the next layer, we will have 5*5*3 by 32*32 connections which are 76,800 connections. When compared to a fully connected layer (3 million connections), the number of parameters which must be learned is significantly reduced. As a result, employing the CNN network can drastically reduce the number of parameters required [15]. Fig. 3.12 demonstrates a typical neural net with just one hidden layer (a): a fully connected architecture wherein every input neuron is connected to all the neurons of the hidden layer. In contrast, we have for (b): a convolutional architecture which reduces the number of weights. A convolution kernel size of 2 moves over the input layer and projects the output to the next neuron of the hidden layer. The convolutional neural net has vast applications in both image and signal processing. Different convolutional neural net architectures have been introduced for purposes such as image or signal denoising [1,19-22].

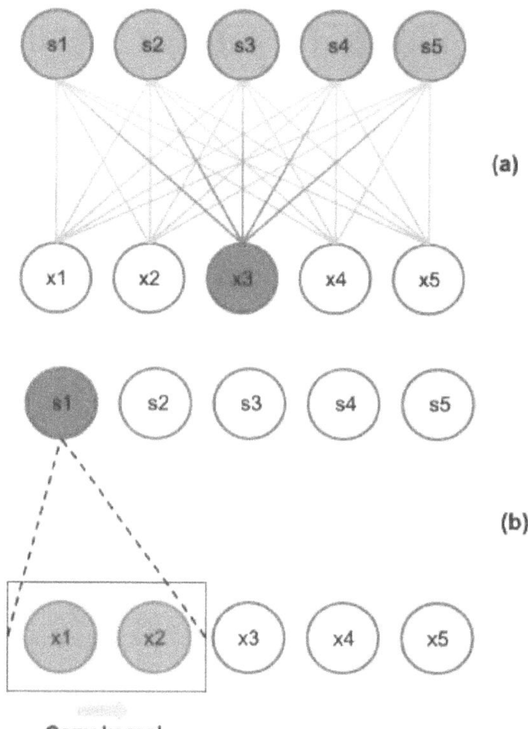

Fig. 3.12. (a) represents typical neural net that uses fully connected architecture. All input nodes (x) connected to the next layer nodes by some weights. For instance, input x1 is connected to the next layer by five weights. (b) demonstrated a convolutional architecture neural network. A convolutional kernel with a kernel size of two is moving over the input and, by applying the convolution operator, sends the output to the next layer. Implementing a convolutional layer decreases the required number of weights. Instead of five weights in (a), applying a convolutional layer reduces that to just two weights.

3.13.3 The convolution operation

In its most general form, convolution is a mathematical operation performed on two functions of a real-valued argument. The convolution of two functions x and w is defined as [1]:

$$s(t) = \int x(a)w(t - a)da \qquad (4.21)$$

The first argument is commonly referred to as the input and the second as the kernel in convolutional neural network terminology [1]. Usually when we work with data on a computer, we employ discretized data. The discrete convolution is defined as [1]:

$$s(t) = \sum_{a=-\infty}^{\infty} x(a)w(t - a) \qquad (4.22)$$

The input dataset is commonly a multidimensional array of data in machine learning applications, like RGB images and hyperspectral images. Here the kernel is usually a multidimensional array of parameters which are adjusted by the learning algorithm [1].

3.13.4 Convolutional neural networks assumptions

Convolution is based on three key ideas which can aid in the improvement of a machine learning system. These key ideas are: sparse interactions, parameter sharing, and equivariant representation [1]. Matrix multiplication by a matrix of parameters is used in traditional neural network layers (not convolutional), with a separate parameter defining the interaction of each input unit and each output unit [1]. As a result, each output unit interacts with each input unit, resulting in a high computational cost. However, the interactions in a convolutional neural network are usually sparse (sparse weights). In sparse connectivity kernel size is chosen to be smaller than input size, Fig. 3.12(b). Using the same parameters for multiple functions in a model is referred to as parameter sharing [1]. When computing the output of a layer in a classic neural network, each weight matrix element is used exactly once. It's multiplied by a component of the input and then never used again [1]. The convolutional neural network, on the other hand, employs each member of the kernel at all input points (except at some of the boundaries, like edges in images). Because of the parameter sharing employed by the convolution operation, we have to learn only one set of parameters rather than a different set for each location [1]. Convolution is therefore far more memory and statistically efficient than is dense matrix multiplication [1].

In a convolutional architecture, parameter sharing offers the layer a property called equivariance to translation [1]. Translational equivariance, or simply equivariance, is a critical property of convolutional neural networks in which the location of an object within an image should not be fixed in order for the CNN to detect it. This means that, if the input changes, the output will change as well. When processing images, moving the input one pixel to the right will cause the representations to move one pixel to the right as well. The idea of weight sharing is used in CNNs to obtain the property of translational equivariance. Because the same weights are applied to all images, if an object appears in any of them, it will be detected regardless of its location. This attribute is useful for image classification and object detection in situations where the object may appear several times or may be in motion. This feature can also be used for signal processing [1].

3.13.5 Autoencoder neural network

An autoencoder neural network is a type of unsupervised learning algorithm which is trained to map input to output. The network may be viewed as consisting of two parts: (i) an encoder function $h = f(x)$ and (ii) a decoder function which produces a reconstruction $r = g(h)$ where x is the input data and r is the network's output. The identity function, which can transfer input to output exactly, appears to be a simple function; nevertheless, by imposing limitations on the network, such as restricting the number of hidden layer units, we can uncover the fundamental structure of the input data [1,16]. Fig. 3.13 shows a typical autoencoder architecture with one hidden layer. Input layer nodes map into the hidden layer (red circles) which is called latent space.

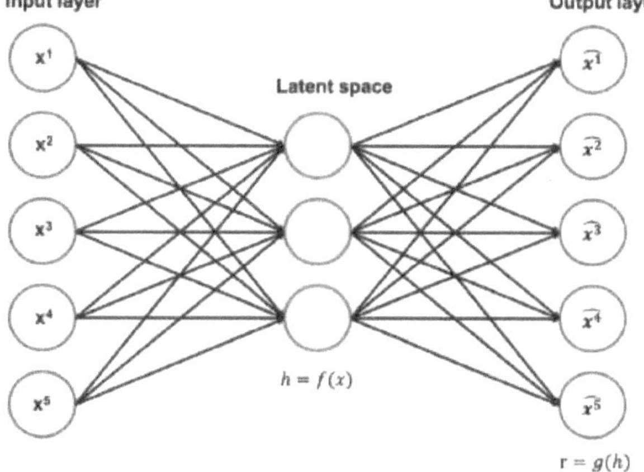

Fig. 3.13. Conceptual drawing of a simple Neural Network Autoencoder. Input layer is connected to the latent space with weights (5 weights for each node). The Latent Space (red circles) is the bottleneck which contains the compressed representation of the input data. Latent space is also connected to the output layer by some weights. By using an iterative approach with backpropagation, model tries to reconstruct the input signal. Mathematically it can be shown that $h = f(x)$, which h is mapped to the latent space and $r = g(h)$,that r is output of the model. The goal of the training is to reconstruction of the input.

Autoencoders are a type of feedforward neural network which may be trained using the same approaches as for feedforward neural networks, such as mini-batch gradient descent with back propagation gradients. While copying the input to the output may appear to be pointless, the encoder's output is usually most beneficial. We expect that training the autoencoder to execute input copying will result in the extraction of valuable input data structure. In other words, it can be viewed of as a dimensional reduction technique wich allows the input dataset to be represented with fewer features. The input dataset will be reproduced using extracted essential features which are mapped in the latent space [1].

Limiting the encoder function f to have a smaller dimension than the input data is one technique for getting useful features from the autoencoder [1]. Undercomplete autoencoders have an encoder dimension smaller than the input dimension. The autoencoder is forced to capture the essential features of the training data by learning an incomplete representation. The learning process is described simply as minimizing a loss function:

$$L\left(x, g\big(f(x)\big)\right) \tag{4.23}$$

where L is defined as the loss function. Minimization of the loss function is proportional to decreasing the difference between input and output of the model. By defining the decoder g as a linear mapping function and the loss function as the mean squared error, an undercomplete autoencoder would perform as a PCA algorithm. However, by defining both encoder and decoder to be nonlinear functions, the model would learn more about the dataset than just copying the input to output [1].

3.13.6 Denoising Autoencoder

An autoencoder with a high-capacity encoder and decoder fails to learn anything valuable (i.e. no reduction) about the input data. A similar issue arises if the latent space has a dimension equal to or greater than the input [1]. In a perfect scenario, an autoencoder architecture may be effectively trained by selecting the latent space dimension and model capacity based on the complexity of the distribution to be modelled [1]. Regularized autoencoders are one solution to this problem. Regularized autoencoders utilize a modified loss function which encourages the model to have characteristics other than replicatiion of input to output. This is in contrast to limiting the model capacity by forcing the encoder and decoder to be shallow (fewer layers) [1]. Another technique is a denoising autoencoder which we have used here to extract useful information from the input dataset for denoising and classification purposes. Rather than modifying the cost function (regularized autoencoder), we can modify the cost function's reconstruction error term in order to develop an autoencoder which learns valuable information about the input data. Typically, autoencoders minimize some loss function 4.23 however, a denoising autoencoder minimizes:

$$L\left(x, g\big(f(\tilde{x})\big)\right). \tag{4.24}$$

where $\tilde{x}$ is the noisy version of x. Instead of just copying their input, denoising autoencoders must first denoise the input x. Denoising processes force both encoder and decoder components to learn the structure of the input data distribution $p_{data}(x)$ [1,17-18]. As a result, denoising autoencoders are another example of how valuable properties can appear as a result of reducing reconstruction error.

The denoising autoencoder is an autoencoder trained to recover the original, uncorrupted data as its output, upon receiving corrupted input data [1]. A denoising autoencoder's training procedure is depicted below. A corruption process $C(\tilde{x}, x)$ is introduced which adds noise to the original uncorrupted dataset. During training, autoencoder learns a reconstruction distribution $p_{reconstruct}(x|\tilde{x},)$ by seeing examples of both original (uncorrupted version) and corrupted (noisy version) data [1]. This procedure is described as follows [1]:

1- Sample a training example x from the training dataset.
2- Add noise to the training examples and then sample from that ($\tilde{x}$).
3- Use $(x, \tilde{x})$ as training example. Train the autoencoder with the task of reproducing an uncorrupted x by comparing both $\tilde{x}$ and x.
4- Feed the trained autoencoder with corrupted version data ($\tilde{x}$) and extract the uncorrupted version (x) as the autoencoder output.

3.13.7 Convolutional autoencoder

Convolutional autoencoders are built on a typical autoencoder design which includes convolutional encoding and decoding layers [19]. Convolutional autoencoders are better suited to image processing than are conventional autoencoders because they use the full power of convolutional neural networks properties. They are also suitable for signal processing purposes, such as EGG signals [20-21].

3.14 Convolutional autoencoder in coherent Raman microcopy

Implementing a convolutional autoencoder network was our first goal in overcoming the noise problem in hyperspectral SRS imaging. The autoencoder neural network compresses the input data dimension to a low dimension, as discussed above. The encoder layer extracts useful information from the input data and maps this onto the latent space. A well-trained autoencoder can reconstruct its input data set with only a few essential data features mapped to the model's latent space, achieved by up-sampling with the decoder component of the model. We take advantage of this dimension reduction for image segmentation and material classification in the sample. The features used for input data reconstruction were also used to distinguish different materials in the sample. We also demonstrated that the autoencoder topography can be used for other microscopy techniques such as pump-probe microscopy.

In the next chapter, we introduce two customized convolutional autoencoder neural networks named UHRED and SHRED. The difference between these is the way in which the input dataset is feed into them. The SHRED is based on supervised learning, whereas the UHRED is an unsupervised learning model. We trained these two models on a synthetic dataset in order to gauge their performance. Subsequently, a real SRS microscopy data set was used to demonstrate their capabilities for image denoising and classification.

References

[1] Ian Goodfellow, Yoshua Bengio, and Aaron Courville. Deep Learning. 2016. MIT Press

[2] A. Samuel, "Some Studies in Machine Learning Using the Game of Checkers," IBM journal, 1959.

[3] T. Mitchell, "Introduction to machine learning," Machine learning, vol. 7, pp. 2-5, 1997.

[4] A. Burkov, The hundred-page machine learning book, 2019.

[5] Svante Wold, Kim Esbensen, Paul Geladi, Principal component analysis, Chemometrics and Intelligent Laboratory Systems, Volume 2, Issues 1–3, 1987, Pages 37-52,

[6] J. MacQueen, "Some methods for classification and analysis of multivariate observations," in Proceedings of the Fifth Berkeley Symposium on Mathematical Statistics and Probability, 1967.

[7] Pascal Vincent, Hugo Larochelle, Isabelle Lajoie, Yoshua Bengio, and Pierre-Antoine Manzagol. Stacked denoising autoencoders: Learning useful representations in a deep network with a local denoising criterion. The Journal of Machine Learning Research, 9999:3371–3408, 2010.

[8] V. N. Vapnik, "An overview of statistical learning theory," IEEE Transactions on Neural Networks , vol. 10, no. 5, pp. 988-999, 1999.

[9] V. N. Vapnik, The nature of statistical learning theory, 1995.

[10] S. Watanabe, Algebraic geometry and statistical learning theory, Cambridge university press , 2009.

[11] P. L. B. Martin Anthony, Neural network learning: Theoretical foundations, Cambridge University Press, 2009.

[12] S.-C. Wang, Interdisciplinary Computing in Java Programming Language, Springer Science+Business Media New York, 2003.

[13] Y. B. ,. G. H. Yann LeCun, "Deep learning," Nature, no. 521, pp. 436-444, 2015.

[14] Rumelhart, D., Hinton, G. & Williams, R. Learning representations by back-propagating errors. *Nature* **323,** 533–536 (1986)

[15] S. Albawi, T. A. Mohammed and S. Al-Zawi, "Understanding of a convolutional neural network," *2017 International Conference on Engineering and Technology (ICET)*, 2017, pp. 1-6

[16] C. Doersch, "Tutorial on Variational Autoencoders," arXiv.org, p. arXiv:1606.05908 , 2021.

[17] Vincent, P., Larochelle, H., Bengio, Y., Manzagol, P.A.: Extracting and composing robust features with denoising autoencoders. In: ICML, pp. 1096–1103 (2008)

[18] Junhua Li, Zbigniew Struzik, Liqing Zhang, Andrzej Cichocki, Feature learning from incomplete EEG with denoising autoencoder, Neurocomputing, Volume 165, 2015, Pages 23-31,

[19] Masci J., Meier U., Cireşan D., Schmidhuber J. (2011) Stacked Convolutional Auto-Encoders for Hierarchical Feature Extraction. In: Honkela T., Duch W., Girolami M., Kaski S. (eds) Artificial Neural Networks and Machine Learning – ICANN 2011. ICANN 2011. Lecture Notes in Computer Science, vol 6791. Springer, Berlin, Heidelberg.

[20] L. Gondara, "Medical Image Denoising Using Convolutional Denoising Autoencoders," in IEEE, Barcelona, 2016.

[21] B. Du, W. Xiong, J. Wu, L. Zhang, L. Zhang and D. Tao, "Stacked Convolutional Denoising Auto-Encoders for Feature Representation," in *IEEE Transactions on Cybernetics*, vol. 47, no. 4, pp. 1017-1027, April 2017,

Chapter 4

Application of deep learning in nonlinear optical microscopy

4. Introduction

As discussed in the first two chapters, Coherent Raman microscopy (CRM) is a powerful nonlinear optical imaging technique based on contrast via Raman active molecular vibrations. CRM offers rapid, sensitive, chemical-specific and label-free 3D sectioning of samples and has been applied to fields ranging from biology, to medicine, to mineralogy [1-6]. Although acquisition speeds at up to video-rate have been developed for imaging at a single (fixed) Raman shift, the spectral overlap of Raman bands and the presence of non-resonant background signals can challenge attribution to specific molecular species. Therefore, multiplex or hyperspectral Raman imaging modalities [4, 7-10] were developed to address this issue. Broadband hyperspectral Raman imaging often takes advantage of synchronized fs oscillators, optical parametric oscillators (OPO's), and/or supercontinuum generation in Photonic Crystal Fibres (PCFs). Spectral focusing [10-19] makes it possible to obtain high Raman spectral resolution from such broadband sources. These advances resulted in improved CRM imaging and, of relevance here, much larger data volumes. Strategies proposed to analyze hyperspectral datasets include multivariable curve resolution [20-21], independent component analysis [22], vertex component analysis [23] and the k-means clustering algorithm [24]. In the CRM context, these methods assign a meaningful label to each pixel based on Raman vibrational spectroscopy, helping to classify materials in the sample. Here we present a new and, importantly, unsupervised approach to Machine Learning denoising and classification for which only a single field-of-view, low signal-to-noise ratio image is required for training. We anticipate that this "one shot" approach will help broaden the applications of Deep Learning to hyperspectral optical and nonlinear optical microscopy.

A common issue afflicting CRM is low signal-to-noise (SNR) ratios which can originate from, as examples, low concentrations of target molecules in the sample, and/or scattering losses in deep-tissue imaging [25-26]. Furthermore, in some CRM imaging applications (e.g., *in vivo*), fast imaging and/or low input laser power is required to prevent sample photodamage, typically resulting in low contrast (low SNR) images. Low SNR images also typically suffer from poorly resolved spectral features. To date, various denoising algorithms were applied to image enhancement. However, these typically require either prior knowledge of the noise source or multiple images of the same field of view (under better observation conditions) to enable averaging, which may result in reduced spatial-spectral resolution in the image [27-28] or even sample photodamage. A central issue in SRS microscopy is sample segmentation: translating from a 2D hyperspectral image to a chemical concentration map. In heterogeneous samples, Raman vibrational bands likely overlap and their unmixing and subsequent segmentation typically requires chemometrics methods [20-24].

Deep learning has emerged as a powerful and general tool for data analysis. Deep learning models have demonstrated remarkable performance in imaging applications such as super-resolution microscopy and cancer detection, amongst others [29-35]. Applications of deep learning and machine learning models in CRM have resulted in, for example, advances in lung cancer diagnosis [35], the analysis of expressed human meibum [36] and sample classification [37]. The well-known U-net architecture [38] is one example of deep learning applied to denoising SRS microscopy images [39]. Indeed, the U-net architecture demonstrated excellent denoising of SRS images but required access to both high-SNR and low-SNR image data during model training, necessitating 40 fields-of-view over an extensive (7 hour) training period. While obtaining such data is possible, for cases of delicate or rare biological samples or for the tracking of transient events (requiring high frame-rate imaging), extended training may be more challenging or even impossible. In a novel implementation, U-net was used for denoising Raman spectra in a supervised manner [40]. A powerful customized architecture named DeepChem was very successfully applied to SRS microscopy for material classification and fast imaging [41]. DeepChem uses spectrally summed hyperspectral SRS images for training, potentially resulting in some loss of frequency-dependent information and requires an image augmentation technique in its training.

We consider differences between various deep learning-based methodologies in terms of their requirements for training data. Deep learning models may be classified into two types: (1) supervised and (2) unsupervised. Supervised learning requires labelled training data: all previously reported applications of deep learning to CRM belong to this class. In contrast, unsupervised deep learning is a branch of deep learning where input data is "unlabeled." Such data tends to be much easier to obtain as they require no human intervention for labelling. However, these can be more challenging to work with because the training signals are limited. An autoencoder is a common type of unsupervised machine learning neural network topography. Autoencoders are based on the concept of an information bottleneck which forces the network to compress a signal into a low dimensional space. A significant capability here is the ability to project high-dimensional data into a low dimensional space based on the correlations and similarities observed within an unsupervised, unlabelled training set. It is an example of feature (or representation) learning.

4.1 Chapter goals

Here we present a network topography and training procedure which overcomes previous limitations while maintaining the advantages of neural network-based image processing. Our new approach, UHRED (Unsupervised Hyperspectral Resolution Enhancement and De-noising) works in a label-free manner and has no requirement for high SNR "ground truth data". Moreover, as we demonstrate here, UHRED can automatically segment and label distinct features within a dataset without human oversight. For our demonstration, we used both synthetic and experimental datasets. The synthetic dataset generated by mixing some Gaussian shape signals and feeding them into a 3D matrix in order to make a hyperspectral image (data cube). For experimental dataset, we recorded hyperspectral nonlinear optical spectral signals (termed here the hyperspectral index), in a real sample, at each image pixel. Importantly, the hyperspectral index is very general and could represent any linear or nonlinear optical signal channel including SRS, Pump-Probe, thermal lensing and cross-phase modulation microscopy or any other signal types which depend on an input laser parameter. For the case illustrated here, the hyperspectral index is the SRS vibrational Raman spectrum. Importantly, automatically denoised and segmented images are more easily interpretable by non-experts, thus expanding the range of applications of deep learning in CRM. UHRED is a label-free, unsupervised deep learning approach which: (i) permits reduction in either the optical power requirements or integration time needed to achieve a target SNR; (ii) can work with training data collected only under low-SNR conditions (requiring no high-SNR data); (iii) offers unsupervised identification and segmentation of spectrally distinct materials based on a full (multimodal) hyperspectral data set. We note that previous work in this field has relied on the availability of either high SNR or 'hand-labelled' data. This may be considered a restriction, as data acquisition and labelling are often costly, and may restrict such image processing to users with extensive domain expertise. Conversely, UHRED is a fully automated and unsupervised approach which can be straightforwardly applied and interpreted by non-specialists. As we demonstrate below in an ore sample from Lithium minerals processing, the UHRED method automatically improves contrast and Raman spectral resolution and offers complete segmentation to produce chemical (mineral species) maps.

4.2   Deep neural net architecture

To demonstrate the application of convolutional autoencoders (CAE) to the denoising and segmentation of hyperspectral images, a 10-layer neural net architecture was designed. The architecture of the model is shown in Fig. 4.1 and discussed below. The encoding module had four 1D convolutional layers which followed by a fully connected layer. Convolutional layers consist of an array of kernels with channels which operate on the input signal. The kernel size of each convolutional layer and number of nodes in the fully connected layer determines the number of parameters which must be optimized during the training process. Channels at each convolutional layer learn different length-scale features of its input signal. The first layers learn the input dataset's (here signal) overall structures. The deeper layers learn delicate structures, similar to training a deep learning model for image processing tasks. First convolutional layers learn the general aspects of the images like the edges, then fine details of the input images are learned by the deeper layers. A max-pooling layer was applied at the end of each convolutional layer to decrease the input signal dimension. A max-pooling layer is a kernel window with a specific size which moves along the signal and picks the maximum signal within that window. In the encoding phase, each layer projected its extracted signal features onto the next layer. The encoding module was connected to the latent space through a fully connected layer. The decoding module had four deconvolutional layers and one fully connected layer. Each deconvolutional layer applied an up-sampling operator to its input. The upsampling operator, called "Transpose Convolution," was fed from the latent space and then the prior layers tried to reconstruct the input signal over some iterations (backpropagation). Both encoding and decoding modules used the hyperbolic trig function Tanh as a non-linear activation function. In hyperspectral imaging, each pixel contains a full spectrum and therefore every pixel in the image was used as a training, validation and test dataset. The size of the image was 256×256 pixels, which provided us with 65,536 1D spectra. Eighty percent of the spectra were used for training the model, and 20 percent were used as the validation dataset. The model can be trained in two different ways: supervised or unsupervised. The supervised training process compared the ground-truth dataset (high SNR data) and the noisy input dataset (low SNR data) by calculating the mean square error. Using the Adam optimizer [43], the supervised model was trained to reproduce the ground-truth data from noisy input, similar to what was previously reported. The unsupervised model was fed with only the noisy input dataset (low SNR data). Its output was compared with the noisy input dataset by calculating the mean square error. In the unsupervised approach, no ground-truth (high SNR data) dataset was used. Again, the Adam optimizer was used to optimize the network parameters. We name the supervised, trained model as SHRED (Supervised Hyperspectral Resolution Enhancement and DeNoising) and the unsupervised model as UHRED (Unsupervised Hyperspectral Resolution Enhancement and DeNoising). The PyTorch framework was used to build and train our models. The training time was approximately 10 minutes per model using an NVidia K80 GPU. To demonstrate the designed net's unsupervised segmentation ability, the model first trained with an "encoding–decoding" step, explained above. After the network is well trained under its input's reconstruction task, it was used to extract spectral features of the pixels by passing the desired dataset through the encoding module and projecting the result to the latent space. The dimension of the latent space can be regarded as a set of hierarchical feature vectors. To evaluate the effectiveness of the learned features within the proposed net, the traditional 'k-means clustering' algorithm (hereafter, k-means) was implemented on the latent space. Every single pixel was thus assigned to a specific cluster based on its unique spectral features, where the maximum number of clusters required was determined using the elbow method (i.e. determining the inflection point of the variance as a function of k). As an important point of comparison, the well-known method of principle component analysis (PCA) is based on the first two moments of the dataset, whereas k-means makes use of the full distribution (all moments of the distribution).

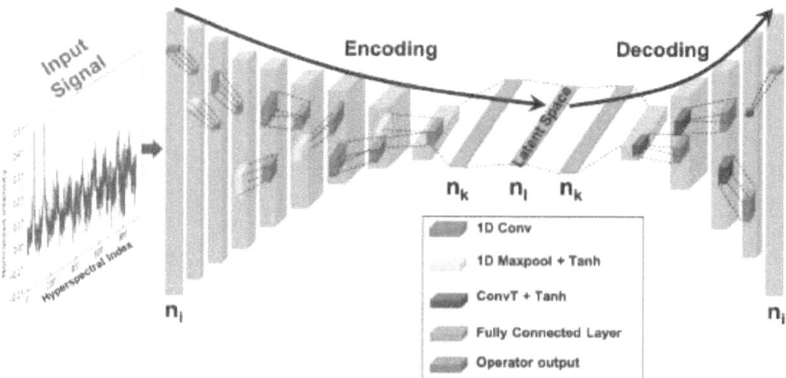

Fig. 4.1. A conceptual depiction of the Convolutional Neural Network Autoencoder implemented here. The Encoding module has four convolutional layers and one fully connected layer. Each convolutional layer (red) applies a 1D convolutional kernel to its input signal, yielding an output (blue) for the next layer to which a 1D Maxpool kernel (yellow) is applied. This cycle is repeated until the last convolutional layer which connects to a fully connected layer (green). The Latent Space (green) is the bottleneck which contains the compressed representation of the input signal. It is connected by another fully connected layer (green) to a Decoding module composed of 4 transpose convolution layers. Each of these up-samples from the preceding layer and, using an iterative approach with backpropagation, tries to reconstruct the input signal. Within the latent space, n_i is the input signal length, n_l is the signal length, and n_o is the output signal length. The full details of the network hyperparameters are reported in the SI.

4.3 Training UHRED model by synthetic dataset

We used five sets of synthetic datasets to train and characterize the proposed model's performance. A mixture of some Gaussian signals with different peak centers, amplitudes, and full-width half maxima was used to generate four types of signals then, feed them into the pixels (generating a hyperspectral image). We first defined four Gaussian signals (see equation 1) and then linear combinations to generate signals more complicated than a simple Gaussian—every generated signal is made up of 94 datapoints. The generated signals are shown in Fig. 4.2a. The generated signals were fed into a zero matrix of the size 256*256*94, making it analogous to a hyperspectral image, here considered as the ground-truth dataset. We could imagine that the four different Gaussian signals correspond to the Raman spectra of four different materials. To make noisy hyperspectral images, we used additive white Gaussian noise at varying ratios of signal-to-noise. In the generated hyperspectral image, each pixel is assigned to either signal or background. To differentiate signal from background, a pixel map image is plotted, shown in Fig 4.2b. We labelled the red, green, orange, blue and gray pixels as materials A, B, C, D and background, respectively.

$$f(\mu,\sigma) = e^{\frac{-(x-\mu)^2}{\sigma^2}}$$ Where μ, σ represent center of the peak and FWHM $\qquad$ (1)

$Signal_1 = f(33,6) + 4f(80,2)$

$Signal_2 = 2.4f(15,1.5) - f(50,5) + f(88,0.5)$

$Signal_3 = 2f(23,4) + f(10,2)$

$Signal_4 = f(45,18)$

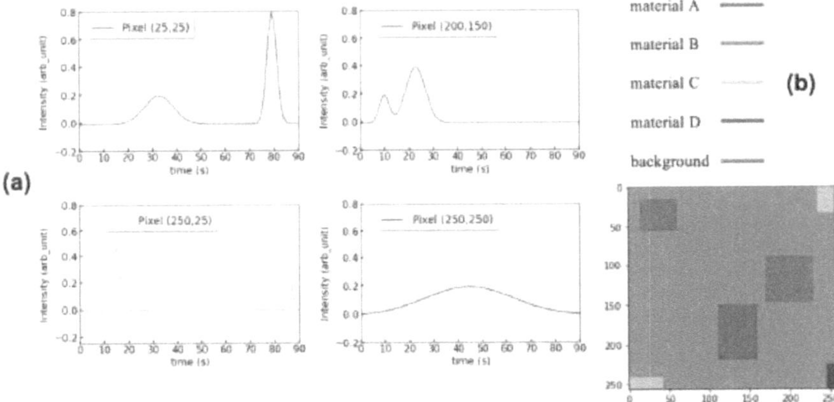

Fig. 4.2. (a): Ground-truth signals. Four signals that generated by mixing different Gaussian signals having different amplitudes, peak centers and FWHM. The red, green, orange, blue and gray pixels respectively belong to material A, B, C, D and background. (b): a pixel map plot that represents the location of each signal.

4.3.1 UHRED model for synthetic data with different SNR values

As explained in the previous section, to train a UHRED model there is no need for a ground-truth dataset. Hence, only a noisy version of the synthetic dataset was used. The training dataset was generated by adding white Gaussian noise to the ground-truth version (Fig. 4.2) so that SNR values of 20, 10, 5 and 2.5 were obtained. The input dataset was then normalized so as to have a mean value of zero and a variance of one. The importance of normalization was explained in chapter four. The noisy input dataset was fed into the UHRED model while the training task's model was to reconstruct its input. Weights in the model were optimized using Adams optimizer and the training time was approximately ten minutes. The training procedure was the same as that discussed in Section 5.1 above. The UHRED model architecture remained constant despite the variations in input SNR values. The trained model's encoder was used to extract helpful input signal features for pattern recognition. The extracted features were then mapped to the latent space. The k-means clustering and the principal component analysis algorithms were implemented in order to assign input signals to specific classes based on their unique signal features. When the SNR was high enough, the clustering algorithms performed very well; however, when SNR became small (SNR = 2.5), the k-means clustering failed to distinguish between different signals and assigning them to their proper classes. The key idea of classifying different species in the sample was based on signal feature differences. The noise suppression capability also has an origin in encoding the input spectra features. The noise features are randomly distributed, making it hard for an autoencoder to learn them all. Thus, autoencoders inherently cannot reconstruct the noise, which thus leads to denoising of a noisy input signal.

4.3.2   Image and signal reconstruction (SNR = 20)

To analyze the trained model's performance, we applied it to a dataset (test dataset) that the model had not seen during training. The test dataset was generated by adding a different white Gaussian noise to the ground-truth data (Fig. 4.2). In hyperspectral imaging, especially for SRS microscopy, noise suppression is one of the essential tasks for samples that required low input laser power or are highly scattering. Fig. 4.3 demonstrates the image enhancement capability of our trained model. In Fig. 4.3, three slices of the noisy hyperspectral image at 'times' 11, 45 and 78 are plotted and compared with ground-truth and the output of the UHRED model. Some materials in the sample which were hardly visible in the input images became apparent after applying the trained model to the noisy input dataset. The image contrast was improved and boundaries between different materials within the sample became clearer. In hyperspectral SRS imaging, species classification in the sample is based on their Raman spectrum. Therefore, we also would like to denoise the Raman spectrum at each pixel. In Fig. 4.4, the capability of the trained model to reconstruct a denoised spectralsignal at every pixel is demonstrated.

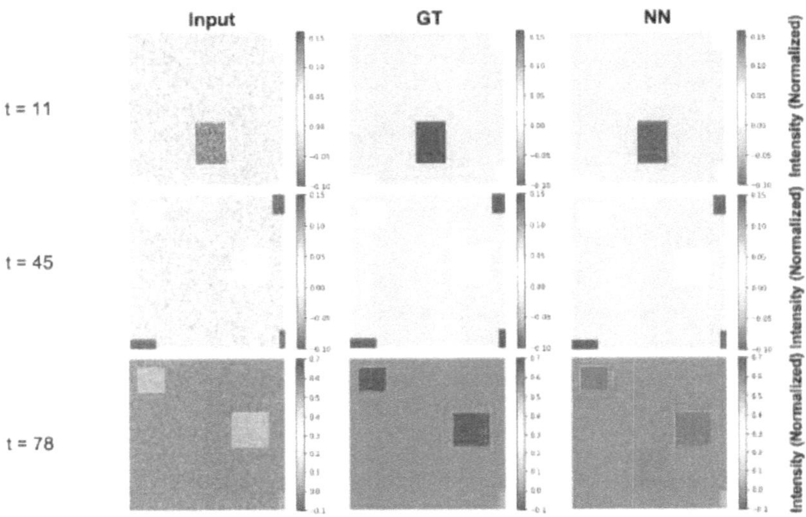

Fig. 4.3. Hyperspectral image reconstruction for three image slices in time. The first, second, and third columns represent the noisy input, ground-truth, and recovered image (the trained neural net's output). Comparing the ground-truth and the NN's output demonstrates the trained UHRED neural net's image enhancement capability. Image contrast significantly improved, which presents the robust capability of a convolutional autoencoder to denoise images.

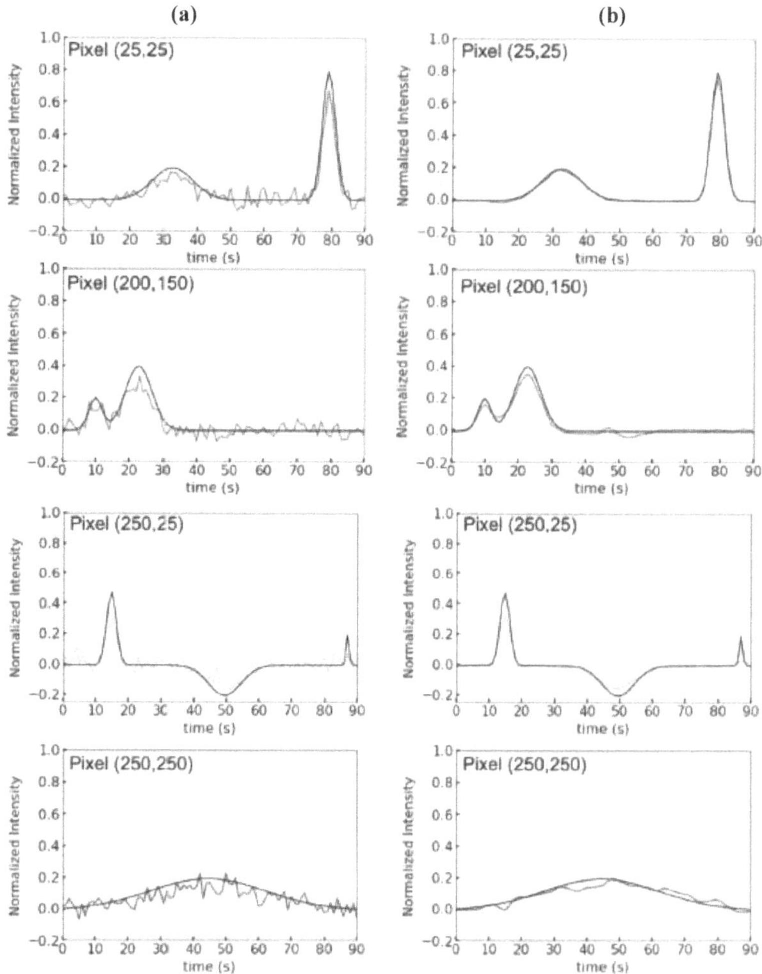

Fig. 4.4. The reconstruction capability of the neural net to denoise and reproduce signal per pixel. The black lines represent the ground-truth signals in both columns. (a) The signals belong to the noisy input dataset, whereas (b) describe reconstructed signals produced by the neural net. The pixels' locations where the signals selected to plot are shown by a number on the top left corner of each plot. Signal denoising is critical because material classification is based on signal features in every pixel for many hyperspectral image modalities like SRS imaging.

4.3.3   Hyper-dimensional clustering algorithm implementation on the latent space

Unsupervised signal classification using a convolutional autoencoder was our second goal. As explained above, a convolutional autoencoder was designed and trained to reconstruct its input. After training the model (UHRED), the encoder was fed with the noisy input dataset. Two hyper-dimensional clustering algorithms, (i) principal component analysis (PCA) and (ii) k-means clustering, were implemented in the latent space. We assumed that the latent space contains unique signal features which could be used to distinguish spectral signals. Here we implemented four convolutional and fully connected layers during the training in order to extract signal features.

4.3.3.1 PCA

Feeding the input dataset to the trained model's encoder resulted in dimensional reduction of the input dataset. The input signals of the dataset consisted of 94 datapoints which the encoder reduced to 20 datapoints. The PCA algorithm was then implemented on the latent space comprised of 20 datapoints. We used PCA with two and three principal components, shown in Fig. 4.5 and Fig. 4.6 respectively. These demonstrate that the convolutional encoder of an autoencoder properly extracts signal features. It can be seen in Fig. 4.5 that the input dataset (65,536 signals), composed of five classes (four different signals and one background), is well recognized by two principal component analysis. The classes are well separated in the reduced dimension space, demonstrating the efficiency of the encoder module for extracting distinct signal features.

Fig. 4.5. We implemented the PCA algorithm to the latent space. PCA with two principal components used here. Five different clusters recognized here are consistent with our pre-knowledge about the input dataset. The input dataset was made of four different types of signals and a background signal. Every colour belongs to a type of signal and the background. The demonstration is just for seeing feature extraction is working properly and different classes are well separated. The colour scheme here is randomly chosen.

To even better visualize the distinguished signals and background in the latent space, PCA with three principal components was implemented, as shown in Fig. 4.6. Again, the autoencoder performed well.

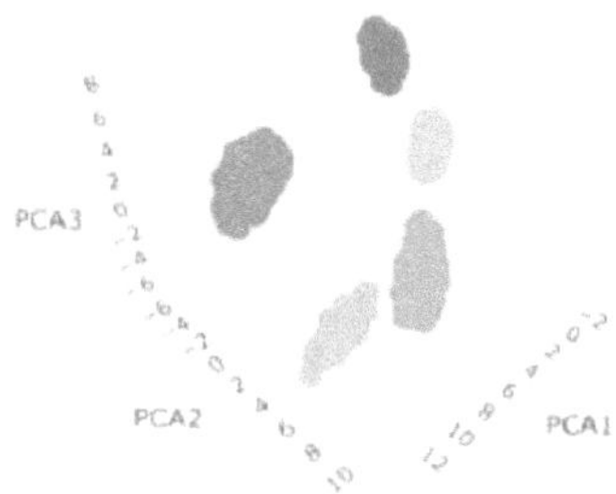

Fig. 4.6. PCA with three principal components implemented on the latent space. As expected, like PCA with two principal components, here also different classes separated well in reduced dimension space, which represents well feature extraction of the model's encoder.

4.3.3.2 Implementing the k-means on the latent space

We considered the latent space to be a feature vector which contains a unique pattern representing each signal. The k-means clustering algorithm was applied to the latent space to classify different sample materials based on their special signal features, similar to what was done using PCA. We used the elbow method to determine the number of centroids in the k-means algorithm. The basic idea of elbow method is that it determines the various cost values as a function of the number of clusters. As the number of clusters increases, there will be fewer elements in the cluster and the average distortion (here measured by the MSE) will decrease. The step where the MSE declines the most is termed the elbow point. In Fig. 4.7. it is seen that three to six clusters are required for the number of centroids chosen in this k-means algorithm. Here we picked five centroids because of our pre-knowledge of the input dataset; however, this could also be done empirically.

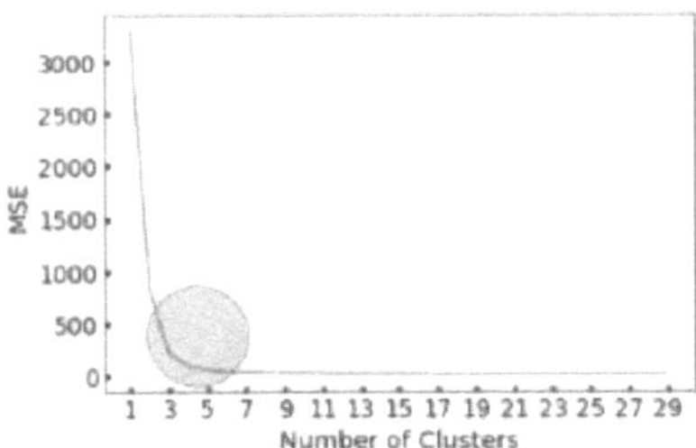

Fig. 4.7. Applying the elbow method on the latent space to determine the best number of clusters in latent space. Between three to six clusters are required, which are considered as the number of centroids in the k-means clustering algorithm.

The k-means algorithm with five centroids was implemented on the latent space in order to classify the input dataset's signals into different classes based on their unique signal features. Applying the k-means algorithm partitioned the latent space to five classes of signals. Here we assigned a colour to each cluster. The result converted into a 2-dimensional matrix of the same size as the input image (256*256) and plotted in Fig. 4.8. The comparison between the pixel map and the clustered latent space demonstrates the autoencoder's ability to extract the input signal features and then use these for classification.

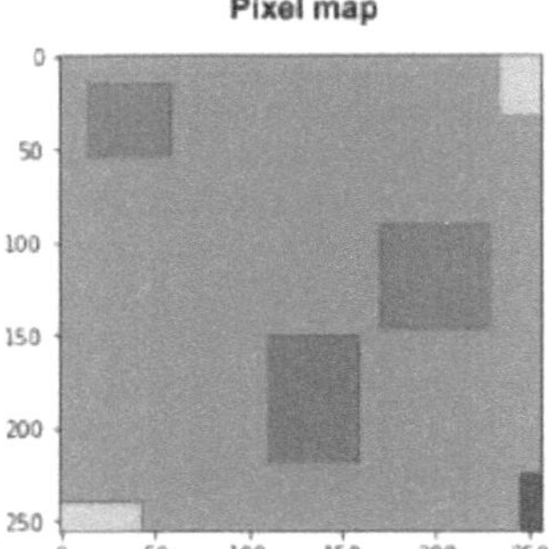

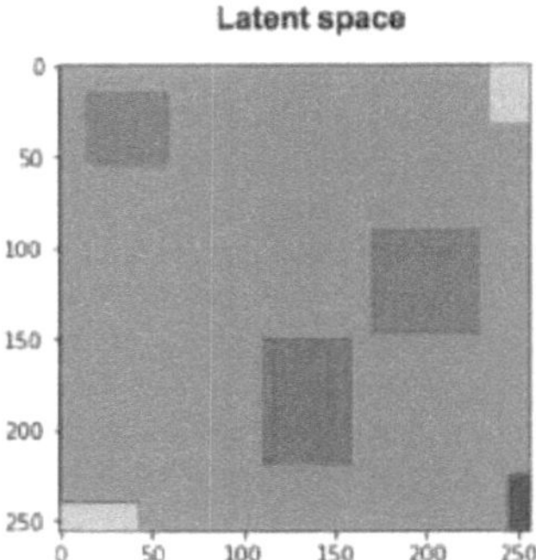

Fig. 4.8. Comparison between clustered latent space by applying the k-means algorithm and pixel map (based on the ground-truth dataset). It demonstrates the perfect performance of the UHRED model in classifying the sample's materials based on signal features. Here the SNR was 20.

4.3.4   Model's performance using noisier input dataset

We also investigated the performance of the model under a different SNR value in the input datasets. In real situations, as in SRS microscopy, it is sometimes necessary to lower the input laser power so as to avoid sample photodamage. This results in the acquisition of low SNR images. Equally, in deep tissue imaging, scattering often leads to low SNR images. We trained three models based on dataset having three different SNR values (in each pixel). The first, second and third datasets had SNR = 10, 5 and 2.5, respectively. By adding noise to the input dataset, the performance of the model decreased, as expected. The model's performance was evaluated in two situations: first, in image/signal recovering ability; second in the classification of different species based on their signal features (Raman spectrum).

4.3.4.1 Model's performance in image and signal reconstruction by varying input dataset SNR value

The signal which was fed into the image pixels to generate a hyperspectral image had the form:

$$Signal = (signal) + B(Random\_White\_Gaussian\_Noise\ (mean = 0, FWHM)$$

where the B is the amplitude of the noise signal. Variation of B resulted in different values of the SNR. The B fixed to values of 0.1, 0.2 and 0.4, equivalent to SNR = 10, 5 and 2.5, respectively. After generating datasets with different SNR values, three models were trained based on the architecture described above. All parameters of the training procedure such as the optimizer's step size, the number of iterations and the optimization algorithm kept the same. The effect of the variation of input SNR on image reconstruction is demonstrated in Fig. 4.9. Although we show here just one signal slice, all reconstructed hyperspectral images behaved similarly.

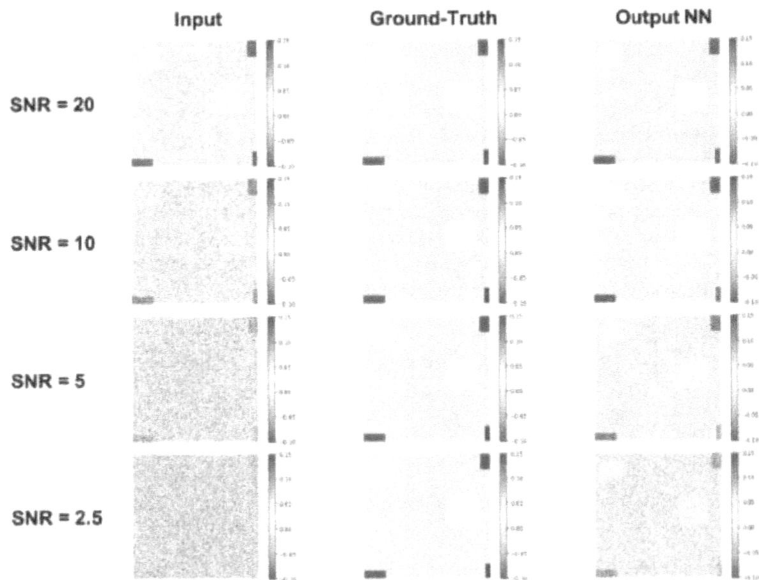

Fig. 4.9. The comparison between the output of different models trained by different SNR input datasets (signals). It demonstrates that decreasing the SNR value leads to a decrease in the contrast of reconstructed images. Images selected at t = 45 from the hyperspectral image. In addition to image contrast reduction, binderies between different materials in the sample also blurred.

The spectral signal reconstruction ability of the models trained by signals with different SNR values was also investigated. In Fig. 4.10. the spectral signal from a pixel at position (25,25) is plotted for the ground-truth, noisy input and the NN outputs. In Fig. 4.10 (a), the ground-truth signal is plotted with its noisy input version, whereas in Fig. 4.10(b), the reconstructed signal is plotted with its ground-truth version. It can be seen that noise suppression performance decreases with decreasing SNR values of the input signal. In the extreme situation where SNR = 2.5, the reconstructed signal does not track the ground-truth signal. This is where our trained model failed to suppress the noise.

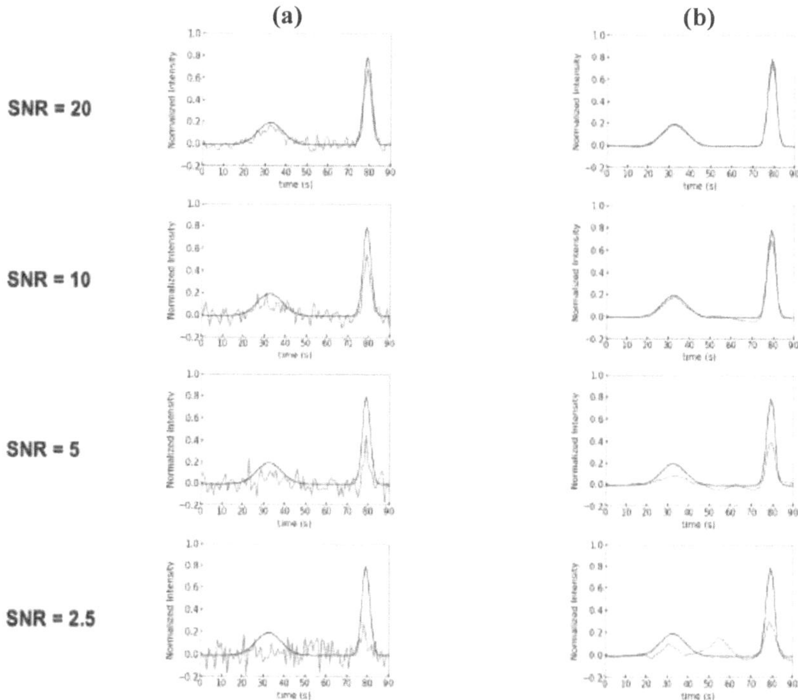

Fig. 4.10. Assessment of the model's signal reconstruction ability by varying the value of SNR for the input dataset. The signal is plotted from a pixel that is located at (25,25). (a) represents ground-truth signal (black) versus noisy input signal (red) with different SNR values. (b) represents reconstructed signal (output of the model) versus ground-truth signal. A higher SNR value results in better signal reconstruction.

4.3.4.2 Implementation of hyper-dimensional clustering algorithms to the latent space

The signal-to-noise variation of the input dataset also influences the performance of the hyper-dimensional clustering algorithms for classification tasks based on signal features. Decreasing the SNR leads to poorer performance of the hyper-dimensional clustering algorithm. By applying the PCA with two principal components, we can see that clusters become proximal, making it difficult to assign a specific cluster to each signal. Furthermore, the k-means algorithm applied to the latent space in the lowest SNR dataset is not as successful as when the SNR is higher. Fig. 4.11 demonstrates the effect of input SNR variation on the PCA algorithm, leading to cluster overlap.

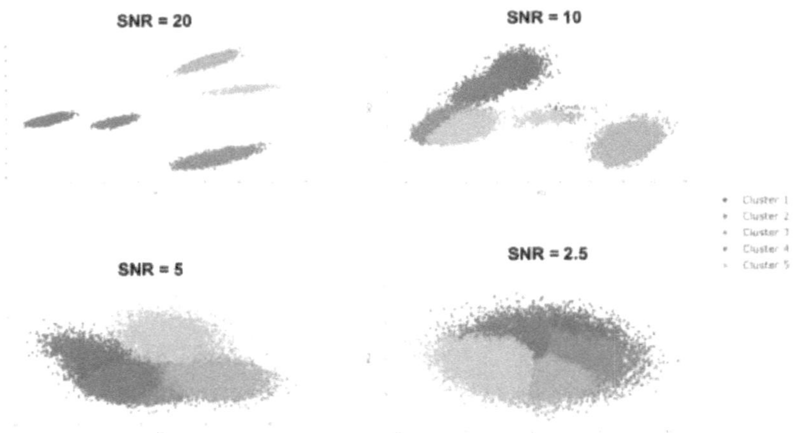

Fig. 4.11 Demonstration of how decreasing the SNR leads to reducing the PCA classification performance. When SNR = 20, different clusters are well separated, whereas, at lower SNR values overlapping between other clusters happens. Here clusters refer to four types of signals in the hyperspectral image and the background.

To visualize the effect on classification of adding noise to the input dataset, we implemented the k-means clustering algorithm on the latent space of four models trained by the input dataset with different SNR values. The pixel map shows signals or background classification, as compared to the clustered latent space (by applying k-means algorithm), is shown in Fig. 4.12. It can be seen that upon decreasing the SNR, the clustered latent space looks less similar to the pixel map, consistent with the results of the PCA algorithm. Low SNR images corrupt crucial features which are essential for image recovery. Thus, the extracted features which should distinguish signals may now distract, resulting in the failure of hyper-dimensional clustering algorithms like the k-means or PCA.

Pixel map

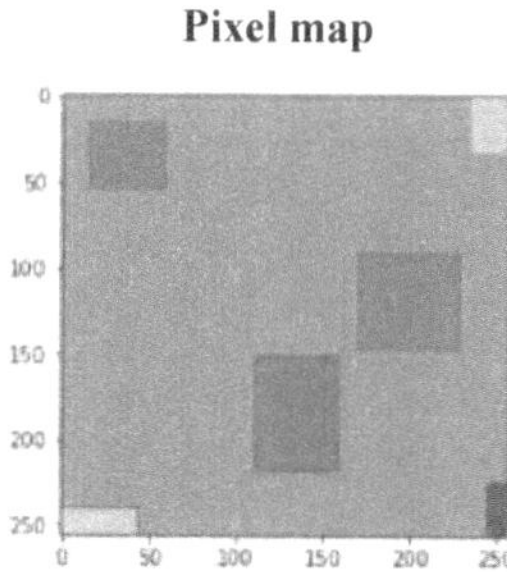

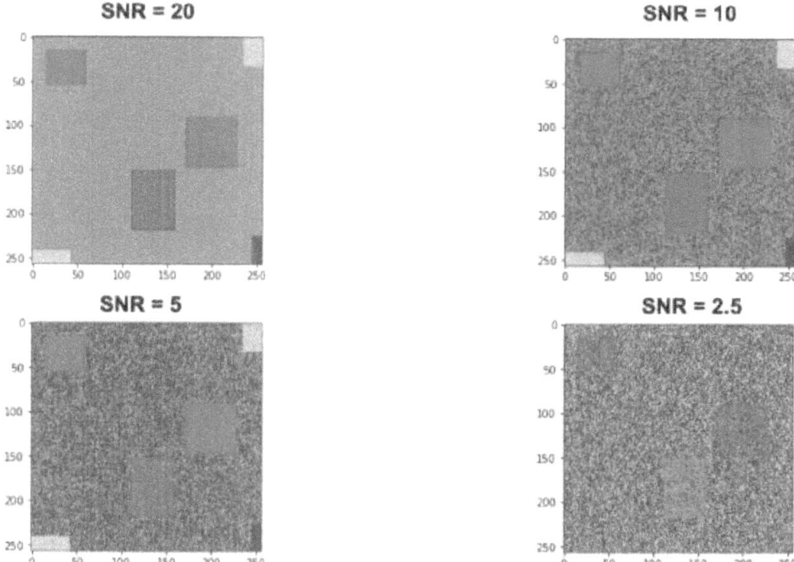

Fig. 4.12. Demonstration of decreasing the SNR value of the input signals leads to lowering the k-means clustering algorithm's performance on latent space. When the SNR value is sufficiently high, like 20 or 10, the clustered latent space looks like the pixel map; however, adding more noise to the signal essential features of the signal are corrupted, which fails the k-means clustering algorithm when the SNR equals to 2.5.

4.4 Training the model with experimental datasets

After our investigation of convolutional autoencoder models using synthetic datasets, described above, we applied our approach to the denoising of experimental hyperspectral SRS images suffering from low SNRs. The trained model was then used to classify different species within the sample based on spectral features which were mapped onto the trained model's latent space. In the following we will describe the experimental setup for hyperspectral SRS imaging. The recorded datasets were used to train three models, one supervised (SHRED) and the two in an unsupervised manner (UHRED). Our input datasets were (i) a heterogeneous binary mixture of hexadecane and distilled water, and (ii) a heterogeneous mixture of three mineral ores from a Lithium mining operation. Both the trained UHRED and SHRED model performed well in denoising the low SNR images and in classifying species within the sample based on their Raman spectrum.

4.4.1 Hyperspectral Stimulated Raman Scattering Microscopy setup

Our spectral focusing hyperspectral SRS imaging arrangement was described previously [10-15]. A schematic of the experimental optical apparatus is illustrated in Fig. 4.13. Briefly, a fs dual output laser system (InSight DS +, Spectra-Physics, USA) produced two synchronized pulse trains at 80 MHz. The fixed wavelength 1040 nm output had a transform-limited pulse duration of 220 fs, with an average power of 1 W. The second output was tunable over the 680-1300 nm range, with a transform-limited pulse duration of approximately 180 fs and a maximum of 1.5 W average power. The 1040 nm was amplitude modulated 1 MHz using a Pockels cell (350-160, Conoptics, USA). Both outputs were linearly chirped by propagation through 60 cm of SF11 glass. Rapid tuning of the Raman resonant frequency was achieved via time-delay scanning of the chirped pulses. The recombined beams were sent to an inverted microscope (IX-71, Olympus, Japan) and focused on the sample with a near IR microscope objective (UPlanSapo, 20x, NA 0.75, Olympus, Japan). Galvanometer mirrors permitted raster scanning across the sample, thus providing a 2D image. After the sample, the forward propagating beams were collected by a second objective acting as the condenser (LUMPlanFI/IR, 40x, NA 0.8 water immersion, Olympus, Japan). A function generator (DS345, Stanford Research Systems, USA) provided the 1 MHz modulation signal for the Pockels cell. The modulation reference signal was sent to a lock-in amplifier (UHFLI, Zurich Instruments) which was used to collect the SRS signal. In this Stimulated Raman Loss (SRL) arrangement, the amplitude modulated Stokes beam was blocked by optical filters (BrightLine 850/310, Semrock, USA and 1064-71 NF, Iridian, Canada), whereas the transmitted Pump beam (containing the modulation transfer) was measured using a large-area photodiode (FDS10X10, Thorlabs, USA). Typically, a few mW of optical power impinged onto the photodiode which was reverse-biased at 50 V. An RF bandpass filter (#3128, KR Electronics) centred at 1 MHz frequency filtered the photodiode electrical output signal. The filtered signal was amplified by a transimpedance amplifier (DHPCA-100, Femto Messtechnik GmbH, Germany), providing the signal input to the lock-in amplifier. A lock-in time constant of 20 µs was used and the relative phase of the lock-in amplifier was adjusted for maximum SRS signal. Data collection and galvo synchronization was performed using ScanImage [42].

In order to normalize the recorded SRS vibrational spectra, determination of the Pump and Stokes wavelengths as a function of time delay was required. This was routinely achieved by recording the sum-frequency generation (SFG) of the instantaneous Pump and Stokes frequencies as a function of their time delay. In the epi-direction, a dichroic mirror (720DCXXR, Chroma, USA) directed back-reflected SFG signals through a short pass filter (750SP, Chroma, USA) to a photomultiplier tube (Hamamatsu H10723-01). KDP powder was used to generate the broadband SFG signals (data not shown) used for *in situ* spectral calibration. Since SFG and SRS are each linear in the Pump and Stokes powers, SFG was used routinely to normalize the spectrally resolved SRS signals from samples.

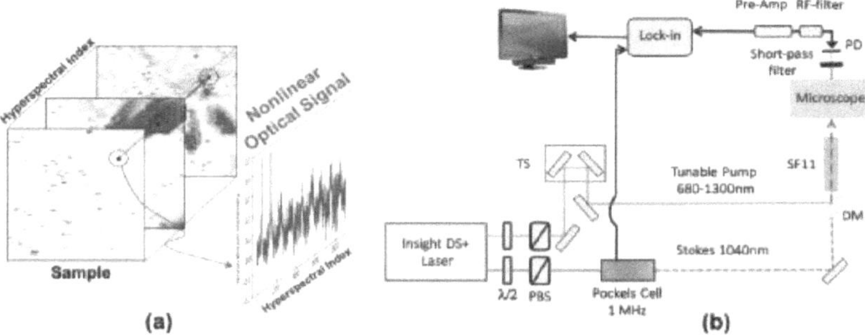

Fig. 4.13. Experimental implementation of hyperspectral nonlinear optical microscopy. (a) Hyperspectral image (data cube) comprised of raster-scanned 2D images (slices) as a function of a selected laser parameter (e.g. frequency, time delay, Raman shift etc.), termed here the hyperspectral index of which only three slices are shown. To illustrate the data cube, we show the variation of a particular nonlinear optical signal at a given pixel (centre of dashed circle). The blue arrow axis indicates the hyperspectral index variation from slice-to-slice. An exemplary nonlinear signal at this single pixel is plotted, in blue, as a function of its hyperspectral index. (b) Hyperspectral Stimulated Raman Scattering (SRS), the hyperspectral index of interest here, as implemented in a nonlinear optical microscope. Depicted are the dual output fs laser system, half wave plate ($\lambda/2$), polarizing beam splitter (PBS), Pockels cell (1 MHz), translation stage (TS), dichroic mirror (DM), highly dispersive glass rod (SF11), microscope, photodiode (PD), RF-filter (bandpass filter at 1 MHz), transimpedance pre-amplifier, and lock-in amplifier. For details, see the text.

4.4.2 Training the models

Two Convolutional Neural Network models were designed and subsequently trained. These were used to demonstrate the utility of convolutional autoencoders for both denoising and segmentation. The first model was used for denoising. It was based on a 10-layer convolutional Denoising Autoencoder net used to denoise a noisy hyperspectral dataset (here recorded at low input laser power). The sample consisted of immiscible hexadecane and distilled water microdroplets. The autoencoder architecture is shown in Fig. 4.1 and the training procedure was discussed in section 4.1 (details of the hyperparameters are given in Appendix 2, Fig. S1). The input dataset (hexadecane and distilled water) size was similar to the synthetic dataset, allowing the same hyperparameters to be used.

The second model, designed to demonstrate unsupervised segmentation and denoising ability, was based on a 10-layer convolutional autoencoder net. This model's architecture is the same as shown in Fig. 4.1 (details of the hyperparameters are given in the Appendix 3, Fig. S2). The model trained with the second sample dataset, which was a complex heterogeneous mixture of three mineral ores from a Lithium mining operation. The dataset is much more complicated than the first. In this second dataset, each pixel contained 909 datapoints, making the spectrum length per pixel ten times longer than the first dataset. Moreover, the second sample (dataset) was composed of more materials within the field of view. This second model was first trained with an "encoding–decoding" step, as described in section 5.1 (UHRED). After the network was trained for its reconstruction task, it was used to extract spectral features within each pixel by passing the desired dataset through the encoding module and projecting the result onto the latent space. The latent space can be regarded as comprised of a set of hierarchical feature vectors which determine that space's dimension. A feature vector is an n-dimensional vector which numerically represents features (e.g. a Raman spectrum) of an object. A traditional k-means clustering algorithm (hereafter, k-means) was implemented in the latent space to evaluate the learned features. K-means partitions a data space into k clusters, each with a mean value. each pixel was assigned to a specific cluster based on its unique spectral features, where the maximum number of clusters required was determined using the elbow method.

4.4.3 Dataset acquisition

We acquired SRS spectra for our training and testing hyperspectral images (data cube), as illustrated in Fig. 4.13(a). Two types of samples were used and, for each, hyperspectral imaging signals (256×256 pixels) were collected in the forward direction and normalized by an independently recorded SFG spectrum in order to correct for spectral power variation in the Pump and Stokes beams. The first type of sample was a heterogeneous mixture of (immiscible) neat hexadecane and distilled water. Hexadecane has a well-known Raman C-H stretch vibrational resonance at 2853 cm^{-1}. To record SRS spectra, the tunable Pump beam was set to 805 nm whereas the Stokes was fixed at 1040 nm, the lock-in time constant was set to 20 µs, and the pixel dwell time was 32 µs. The SRS spectrum was recorded over 94 frames. To train our autoencoder on the hexadecane/water sample, two images of the same field of view but at different input laser powers were recorded. A high SNR hyperspectral image obtained from use of high input laser power (60 mW average power at the microscope input, both Pump and Stokes). A low SNR hyperspectral image obtained from use of 3x lower input laser powers (adding Poisson noise, thus reducing the overall SNR), resulting in noisy Raman spectra at each pixel and concomitant poor image contrast. A second field-of-view, one not seen for training by the autoencoder, was used to test UHRED performance.

A second sample type, used here to demonstrate the application of an Autoencoder for segmentation, was a complex mixture of mineralogical ores (spodumene, feldspar and quartz) from a Lithium mining operation. Spodumene, $LiAlSi_2O_6$, is an important lithium-bearing mineral which has become increasingly important due to the burgeoning electric vehicle industry. Feldspars are aluminosilicate minerals with general formula AT_4O_8 where A is potassium (K), sodium (Na), or calcium (Ca), and T is silicon (Si) and/or aluminum (Al). Quartz is SiO_2. This complex sample presented an SRS imaging challenge, revealing several different SRS bands and other nonlinear optical response peaks (i.e. non-Raman backgrounds). Here, the Pump beam was tuned over the 929-998 nm range and both the Pump and Stokes were adjusted to 70 mW average power at the microscope input. The data acquisition scheme was as above and the spectral scan was acquired over 909 frames.

4.4.4   Image de-noising for the first sample

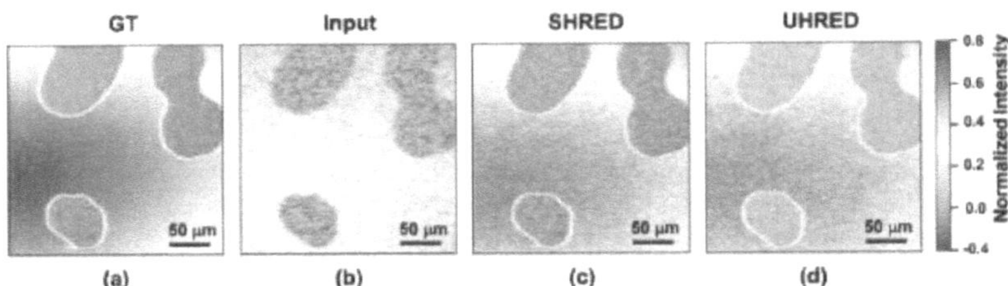

Fig. 4.14 SHRED (Supervised Hyperspectral Resolution Enhancement and De-noising) and UHRED (Unsupervised) methods of image enhancement for hyperspectral SRS of a binary mixture of hexadecane and water, recorded over an invariant field of view (FOV) at (a) high signal-to-noise ratio (ground truth, GT), and (b) low signal-to-noise ratio. The neural net was trained on dataset (a) and then applied to data set (b). In (c) and (d), we show the output of both the SHRED and UHRED models, demonstrating their de-noising and contrast enhancement capabilities. UHRED promises the ability to work with noisier input images and a much smaller training set.

The trained models (both SHRED and UHRED) were used to demonstrate the de-noising and image reconstruction capabilities of our autoencoder nets. An output hyperspectral image was reconstructed by feeding a low SNR hyperspectral image through the trained autoencoder nets. In Fig. 4.14 we show, for a spectral slice at 2853 cm^{-1} Raman shift, hyperspectral images recorded at: (a) high SNR (ground truth, GT), (b) low SNR (noisy Input), and the reconstructed images from both the (c) SHRED and (d) UHRED outputs. It can be seen that both SHRED and UHRED reduced noise and improved image quality, becoming comparable to that of the GT image. We note, however, that in this case the field-of-view was the same for the training set (GT) and the reconstructed images. In order to better assess our autoencoder de-noising capability, we applied the model trained on the GT image of Fig. 4.14(a) to the reconstruction and de-noising of new hyperspectral imaging datasets (recorded at 2853 cm^{-1} Raman shift) to two fields-of-view not seen by the training model, shown in Fig. 4.15 as **FOV 1** and **FOV 2**. In Fig. 4.15(a) we show a noisy input image acquired at low SNR (20 mW). In Fig. 4.15(b) and 4.15(c) we show the UHRED and SHRED images reconstructed from 4.15(a), respectively. In Fig. 4.15(d) we shown the same FOV imaged at high SNR, representing the ground truth (GT) image. Importantly, the GT image was not used here for training and is presented only to permit comparison with the UHRED and SHRED outputs. It can be seen that the autoencoder significantly improves image quality. To further illustrate the autoencoder's de-noising performance, we plot in Fig. 4.15 (e) - the bottom row - the pixel intensity along the dashed line through a ~15 µm droplet within the Region of Interest (ROI) of **FOV 2**. It can be seen that the droplet boundaries are much better defined in the UHRED and SHRED images and are comparable to the GT image.

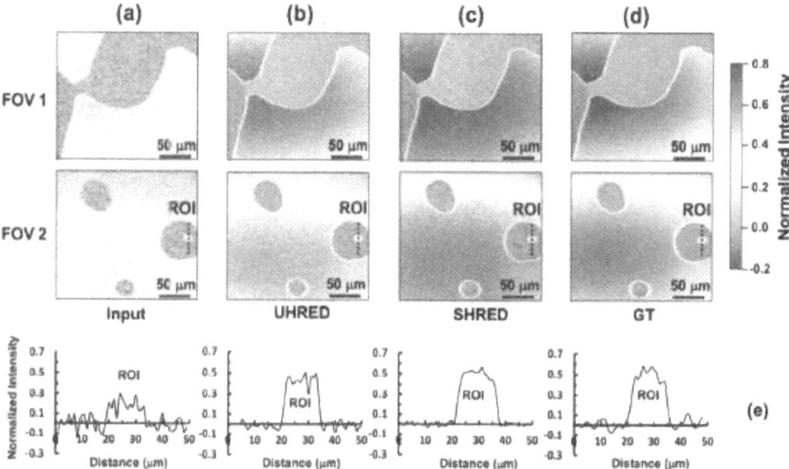

Fig. 4.15 Demonstration of UHRED for SRS microscopy. Hyperspectral SRS imaging of two different fields of view, FOV 1 and FOV 2, shown in the upper and middle rows, for a binary mixture of hexadecane and water. The neural net was trained on the image dataset shown in Fig. 4.14 and then applied here to the FOV 1 and FOV 2 datasets which were not seen by the training model. Column (a) is for a low SNR hyperspectral image. Column (b) shows the UHRED output and column (c) the SHRED output, for both FOVs. Column (d) shows a high SNR hyperspectral image (ground truth, GT) which was not used for training but used here only for comparison with UHRED and SHRED. The bottom row (e) shows signal line-outs along the dashed line through a ~15 μm droplet within the Region of Interest (ROI) of FOV 2. This comparison clearly demonstrates the image de-noising capabilities of both UHRED and SHRED.

4.4.5   Nonlinear optical hyperspectral signal reconstruction

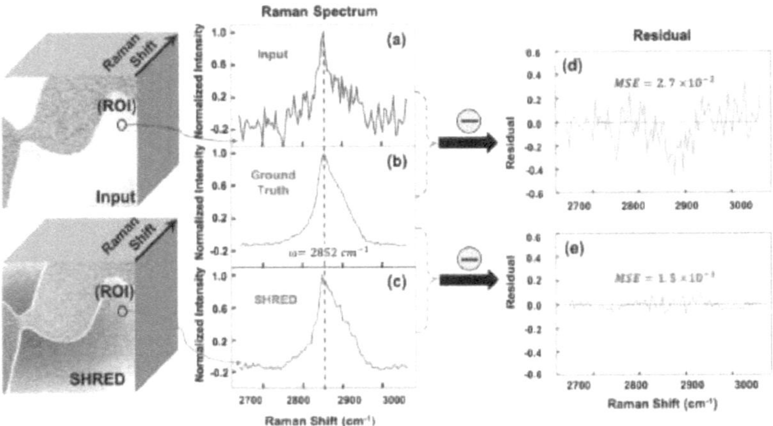

Fig. 4.16 SHRED hyperspectral SRS imaging improves the C-H stretch region Raman spectrum of hexadecane from a single pixel centred within the Region of Interest (ROI). The neural net was trained on a different FOV (not shown) and then applied to the low SNR image (Input, top left), yielding the de-noised output image (SHRED, bottom left). (a) the hexadecane Raman vibrational spectrum (blue) from a single pixel centered in the ROI black circle in the low SNR image (top left). (b) The single pixel Ground Truth (GT) Raman spectrum (green) obtained via high SNR imaging (image not shown) of the same ROI. This GT spectrum was used only for comparison, not training. (c) The single pixel Raman spectrum (red) as extracted by SHRED from the low SNR image (top left), for the same ROI. (d) The difference Raman spectrum (Input – Ground Truth) shows residuals with a Mean Square Error (MSE) of 2.7×10^{-2}. (e) The difference Raman spectrum (SHRED – Ground Truth) shows residuals with a MSE of 1.5×10^{-3}. It can be seen that the SHRED neural net significantly improves (~18x) the SNR in the single pixel Raman spectrum. For details, see the text.

Well-resolved Raman peaks are indispensable for classifying materials within a heterogeneous sample when using a broadband hyperspectral imaging technique such as SRS microscopy. In Fig. 4.16 (top left), we show a low SNR image of a hexadecane-water mixture. In Fig. 4.16 (a) we show the C-H stretch Raman spectrum (blue) of a single pixel centred in the Region of Interest (ROI) shown. In Fig. 4.16 (b) we show a Ground Truth (GT) Raman spectrum (green) recorded at high SNR (image not shown). Importantly, this Raman spectrum was not used for training and is presented here only for comparison. Application of the SHRED net trained on a different field of view (not shown) to the low SNR image (top left), leads to the improved output image (SHRED, bottom left). In Fig. 4.16(c), we show the SHRED output Raman spectrum (red) of the same single pixel centred in the ROI. The neural net improves the Raman spectral SNR at each pixel within the image. To quantify this, we subtracted the GT Raman spectrum from that of the Input and the SHRED outputs. In Fig. 4.16(d), we show the residual Raman spectrum for Input minus Ground Truth which has a Mean Square Error (MSE) (per pixel) of 2.7×10^{-2}. In contrast, in Fig. 4.16(e), we show the residual Raman spectrum for SHRED minus Ground Truth which has a MSE of 1.5×10^{-3}. It can be seen that SHRED improves the SNR in the single pixel Raman spectrum by a factor of ~18. We note that this same analysis applied to the single pixel Raman spectrum for the UHRED net (data not shown) resulted in a poorer residual MSE of 4.5×10^{-3}, but still a factor of 6 improvement over the raw Input signal. It should also be emphasized that the SNR of these (single pixel) Raman spectra could be significantly improved ($\sim \sqrt{N}$) by spatially averaging over N neighbouring pixels.

4.4.6   Signal reconstruction for supervised versus unsupervised models

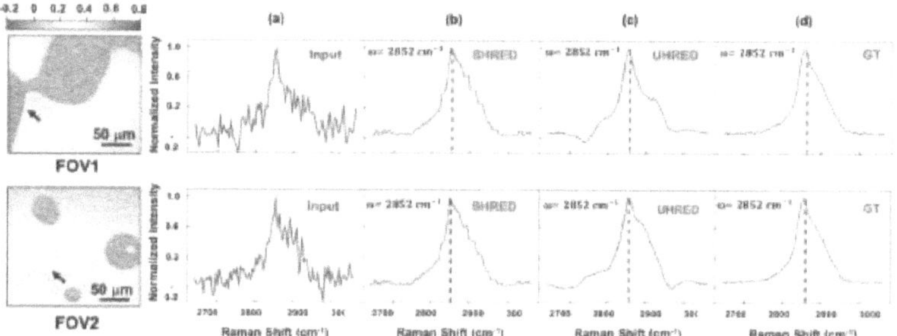

Fig. 4.17 FOV1 and FOV2 recorded at low SNR. (a): Noisy input spectrum of a pixel (indicated by a black arrow), (b) and (c): Output of the SHRED and UHRED model respectively that recovered the same noisy input pixel. (d): Ground truth spectrum of the same noisy input pixel (ground truth FOV is not shown here).

In Fig. 4.17, we directly compare SHRED and UHRED models in terms of their ability to recover single pixel Raman spectra. Here, both SHRED and UHRED are trained on a different field of view, namely that shown in Fig. 4.14. These models were then applied to the low SNR images shown on the left, labelled as FOV1 and FOV2. In Fig. 4.17 (a) we show the low SNR single pixel Raman spectra (blue) of the two pixels indicated by the black arrows in FOV1 and FOV2. The outputs of the SHRED and UHRED nets are shown (red) in Fig. 4.17 (b) and (c), respectively. In Fig. 4.17 (d) we show the Ground Truth (GT) Raman spectrum (green) of the same pixel, recorded at high SNR (image not shown). We note that the GT single pixel Raman spectrum is shown here only for comparison and was not used for training. It can be seen that both models significantly improve the Raman spectral SNR. We note that the output Raman spectrum of the UHRED model is more sensitive to spatial variations of the noise than is the SHRED. Nevertheless, the UHRED output Raman spectrum remains significantly improved over the input and likely suffices to identify key Raman spectral features. We expect that spatial averaging over neighbouring pixels will significantly reduce UHRED sensitivity to the noise spectrum of a given pixel.

4.4.7 Segmentation by an unsupervised Autoencoder

An important second goal of this work is to demonstrate the use of a convolutional autoencoder for unsupervised segmentation. As discussed above, two convolutional autoencoder nets (SHRED and UHRED) were designed and trained. The encoding module of each trained model was fed a low SNR input image dataset, for example from a hexadecane-water sample. We now apply a hyper-dimensional clustering algorithm (k-means clustering) to the latent space, which contains a compressed representation of the input dataset, in order to classify different sample materials based on their unique hyperspectral features (in this case, Raman resonances). In the present case, the latent space was used as a feature vector containing the unique pattern of each hyperspectrum. A critical advantage of using the latent space as a feature vector for classification purposes is that, unlike other ML algorithms [36-37], it requires no labelling of the samples, thus making it an unsupervised classifier. We used the elbow method to determine the number of centroids in the k-means algorithm shown in Fig. 4.18. The elbow method determines various cost values (e.g. variances) as a function of number of clusters, k. As k increases, there will be fewer elements per cluster and the variance will decrease. The point of inflection, where the variance changes the most upon an increment in k, is the elbow point. In Fig. 4.19, we conceptually depict the unsupervised segmentation process. The hyperspectral image data is first passed through the trained encoder (blue) to the latent space (green) (Fig. 4.1). A hyper-dimensional clustering (k-means) algorithm is applied to the latent space and used to classify each image pixel into a specific cluster. As shown in Fig. 4.19, the trained autoencoder unambiguously assigned, in an automatic and unsupervised manner, image pixels to data classes via their hyperspectral features. For the case of hyperspectral SRS, this is equivalent to creating a chemical map wherein sample constituents are uniquely assigned - in a pixel-by-pixel manner - a chemical identity based on Raman vibrational spectroscopy. In the following, we demonstrate this approach to chemical mapping on a more complex Lithium ore sample.

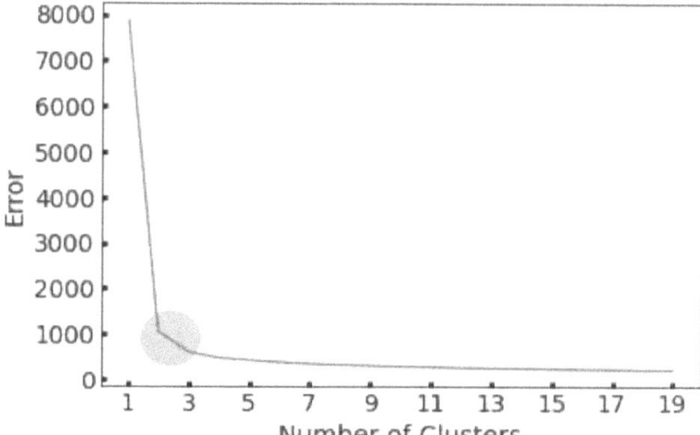

Fig. 4.18 Implementing elbow method to find the proper number of clusters in the hexadecane and water sample. As it was expected two is the best number that refers to hexadecane and water in the sample.

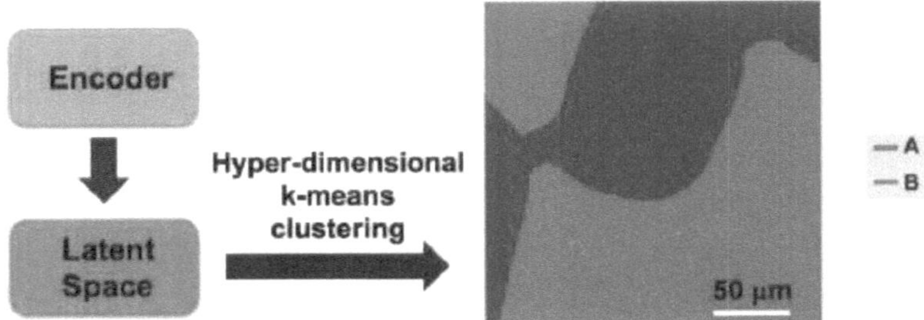

Fig. 4.19 Conceptual depiction of image classification using the trained neural net Autoencoder. The hyperspectral image is passed through the trained Encoder (blue). A hyper-dimensional clustering (k-means) algorithm is applied to the latent space (green). The optimal number of clusters, k, is obtained using the elbow (point of inflection) method. We demonstrate, with a pedagogical example, the application of this classification procedure to hyperspectral SRS imaging data from Fig 4.15(a) FOV1. The k-means procedure automatically and without supervision classifies the image components into unique chemical constituents: in other words, a chemical species map. Here red (A) is hexadecane and blue (B) is water.

4.4.8   Nonlinear optical signal/image reconstruction (mineral sample)

The second UHRED model was designed and trained to demonstrate the autoencoder's unsupervised segmentation of hyperspectral nonlinear optical images. The sample used for this was a complex heterogeneous lithium ore. We recorded hyperspectral nonlinear optical signals (i.e. the hyperspectral index) at each pixel. Although the hyperspectral index is in principle very general, here it is the Pump-Stokes time delay in a spectral focussing scan, directly related to the SRS vibrational Raman spectrum. In Fig. 4.20 (a) we show hyperspectral imaging of Lithium ore samples. At three different single pixels, indicated by the black arrow in each image, we also show the hyperspectral index plots (blue), related to SRS Raman vibrational spectra. These represents the 'noisy' input image data. A UHRED autoencoder was trained on an input hyperspectral data cube from a different Lithium ore sample (not shown). In Fig. 4.20 (b), we show the UHRED output (red) at the three pixels shown in the raw input data in (a). It can be seen that UHRED very significantly improves the SNR of the hyperspectral index at all pixels simultaneously, without any loss of spectral resolution. Importantly, UHRED differs from what may be achieved by other denoising methods such as moving average smoothing, as the latter reduces both the spectral resolution and peak contrast. To directly compare these, we applied a moving average filter with a kernel size of 10 to each pixel in the image. In Fig. 4.20 (a) we show, in red, the moving average spectrum (m_avg) superimposed on each (blue) hyperspectral scan. It can be seen that there is a reduction in both spectral resolution and peak contrast in the moving average spectra as compared to the UHRED spectra in (b). We note that the few green pixels seen in the images are due to detector saturation: these intense but very localized signals originate from other (non-SRS) modulation transfer signals such as thermal lensing, likely due to strongly absorbing semiconductor materials (e.g. pyrite) which appear sparsely in such mineral samples.

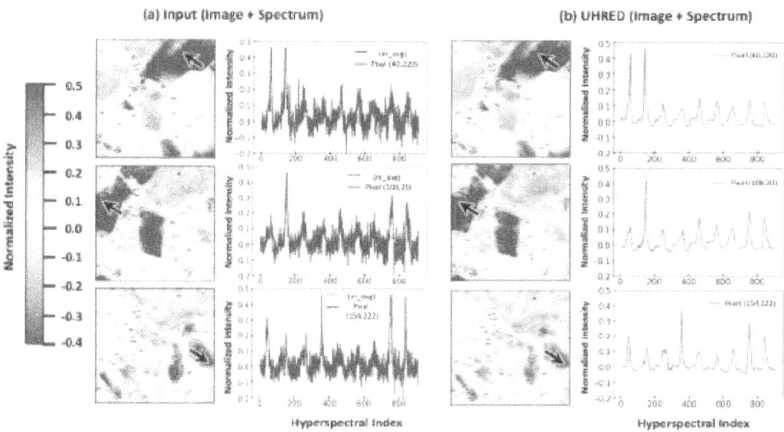

Fig. 4.20 Hyperspectral SRS imaging at three different spectral slice in Lithium ore sample. (a) A hyperspectral image is shown along with a hyperspectral index scan (blue) for the three different pixels indicated by the black arrows in the three panels (top, middle, bottom). Application of a moving average filter with a 10-point kernel is also shown (red) superimposed on each (blue) hyperspectral scan. (b) The output of the trained UHRED Autoencoder applied to the hyperspectral image data of (a). The UHRED output hyperspectral index scans (red) for the same three pixels from (a) are shown (top, middle, bottom). The UHRED improves the hyperspectral index SNR at all pixels simultaneously, without the loss of spectral resolution or peak contrast seen in the moving average filter.

4.4.9   Sample segmentation based on latent space vector (mineral sample)

Material-specific identification and classification (i.e. a chemical species map) using hyperspectral microscopy in a heterogeneous sample is achievable with SRS imaging. However, analyzing such high dimensional data sets typically requires both time and human involvement. We present here an automated, unsupervised approach wherein the UHRED encoder module compresses the high dimensional SRS image features into a low dimensional representation. The compressed representation is mapped to the latent space which is used to distinguish sample constituents based on their hyperspectral index scan (here, the Raman spectrum). A k-means clustering algorithm is then applied to the latent space for pattern recognition. As above, the optimal number of clusters k was determined by the elbow method shown in Fig. 4.21.

In Fig. 4.22 (top), we show the automated image segmentation of a complex Lithium ore sample comprised of quartz (SiO_2, red), feldspars ((K, Na, Ca)$_1$(Si, Al)$_4O_8$, yellow) and spodumene ($LiAlSi_2O_6$, blue). UHRED followed by an unsupervised, automated k-means algorithm, implemented in the latent space, classifies constituents based on their hyperspectral index responses (spectral focussing time delay scans, see Fig. 4.20). Here, we convert the hyperspectral index to SRS Raman shift, permitting direct comparison with the known Raman spectra of these mineral compounds. The randomly scattered green pixels are non-SRS modulation transfer signals saturated at the detector due to strong absorption, discussed above, and are automatically treated by k-means segmentation but are not discussed further here. Below, the Raman spectra extracted by unsupervised UHRED k-means are plotted for each constituent and compared to reference Raman spectra (black dashed lines) of these compounds. It is encouraging that the Raman spectra extracted from the model, in a completely unsupervised and automated manner, are in good agreement with those of the reference compounds. This suggests that the UHRED k-means method presented here is well suited to unsupervised image segmentation. We believe that this will be a useful tool in the routine generation of chemical species maps from imaging data.

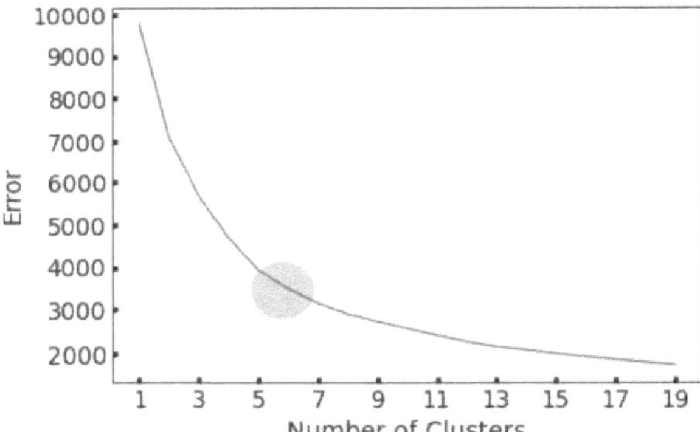

Fig. 4.21 Implementing elbow method on Latent space to find the proper number of clusters in the complex lithium ore sample. As it was expected five is the best number that refers Quartz, Spodumene, Feldspar, hot pixels and background in the sample.

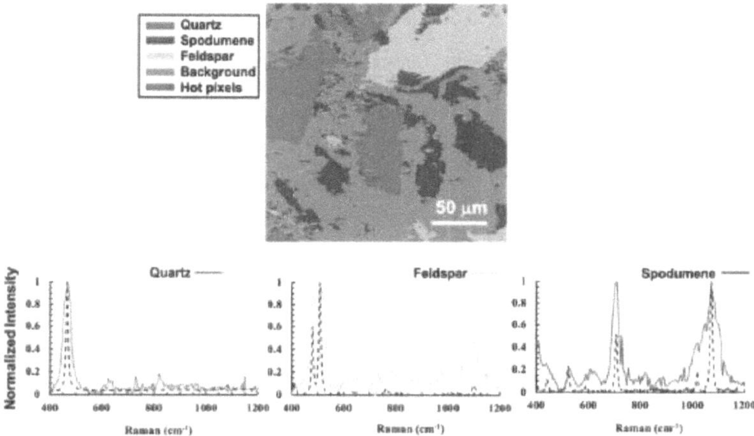

Fig. 4.22 Automated image segmentation of a complex Lithium ore sample (top) comprised of quartz (SiO2, red), feldspars ((K, Na, Ca)1(Si, Al)4O8, yellow) and the Lithium-bearing ore spodumene (LiAlSi$_2$O$_6$, blue). UHRED followed by an unsupervised, automated k-means algorithm, here implemented in the Latent space, classified constituents based on their hyperspectral index responses (see Fig. 4.20). Here, the hyperspectral index is converted to SRS Raman shift to permit direct comparison with the known Raman spectra of these mineral compounds. The Raman spectra (below) extracted by UHRED k-means is plotted for each constituent and compared to the reference Raman spectra (black dashed lines) of these compounds. The general agreement between the extracted Raman spectra and those of the reference compounds permits the UHRED k-means method to segment the image into the chemical species map shown above. For details, see the text.

4.4.10. Limitation

We discuss a limitation of the approach implemented here. Both UHRED and SHRED required a preprocessing step to remove saturated pixels. As noted above, a few pixels in the raw datasets were saturated during data acquisition. To train the models, saturated pixels were replaced by the average value of a randomly chosen subset of nearby pixels. In the current implementation, saturated pixels were identified manually but this process could be automated in future implementations. As the preprocessing phase can be eliminated when a clean dataset is supplied, our technique remains unsupervised.

4.5 Conclusion

We applied unsupervised Deep and Machine Learning techniques to hyperspectral image contrast enhancement and segmentation of chemically distinct species comprising a sample. We introduced two convolutional autoencoders, SHRED and UHRED, with the former being a supervised model comparable to prior approaches, the latter unsupervised and representing a new approach. Both UHRED and SHRED were trained and applied to both synthetic and experimental dataset successfully. The hyperspectral index discussed here is quite general and may represent any laser parameter-based index which is varied. The hyperspectral index scan could therefore represent, as examples, spectral data from SRS, CARS, linear or nonlinear fluorescence, Pump-Probe, Thermal Lensing or Cross-Phase Modulation microscopies. In the examples presented here, the hyperspectral index is the SRS Raman vibrational spectrum obtained in a spectral focussing SRS arrangement. UHRED demonstrably enhances hyperspectral SRS contrast in low SNR images. Furthermore, the extracted single pixel Raman spectra were in good agreement with both the SHRED and Ground Truth spectra (the latter recorded at high SNR). A key point of this work is the unsupervised compression, by an autoencoder, of the high dimensional features of the input dataset to essential latent space features. Autoencoders are based on an information bottleneck which forces compression of a high dimensional data into a low dimensional space based on correlations perceived within an unsupervised, unlabelled training set. The resultant compressed dataset is considered to be a feature vector for each pixel in the SRS image. Importantly, implementing the k-means clustering algorithm to the latent space feature vector provided us with an automated, unsupervised image segmentation procedure. This corresponds to an intuitive chemical species map which is useful in many fields of science and technology. Notably, in the results presented here, only a single unlabelled hyperspectral image was used for the input dataset: larger input datasets would result in even better outcomes. UHRED can be further generalized to include spatial properties of the sample, including birefringence, and to incorporate multimodal signals such as harmonic generation, fluorescence, thermal lensing etc. As computational power continues to increase, we expect that real-time automated, unsupervised image de-noising and material identification will become available to researchers worldwide.

References

[1]. J.-X. Cheng, and X. S. Xie, *Coherent Raman Scattering Microscopy* (CRC Press, 2013).

[2]. J.-X. Cheng, and X. S. Xie, "Vibrational spectroscopic imaging of living systems: An emerging platform for biology and medicine," Science 350, aaa8870 (2015).

[3]. C. Zhang, D. Zhang, and J.-X. Cheng, "Coherent Raman Scattering Microscopy in Biology and Medicine," Annual Review of Biomedical Engineering 17, 415-445 (2015).

[4]. D. Polli, V. Kumar, C. M. Valensise, M. Marangoni, and G. Cerullo, "Broadband Coherent Raman Scattering Microscopy," Laser & Photonics Reviews 12, 1800020 (2018).

[5]. C. Zhang, and J.-X. Cheng, "Perspective: Coherent Raman scattering microscopy, the future is bright," APL Photonics 3, 090901 (2018).

[6]. M. C. Kao, A. F. Pegoraro, D. M. Kingston, A. Stolow, W. C. Kuo, P. H. J. Mercier, A. Gogoi, F. J. Kao, and A. Ridsdale, "Direct mineralogical imaging of economic ore and rock samples with multi-modal nonlinear optical microscopy," Scientific Reports 8 (2018).

[7]. C. H. Camp Jr, and M. T. Cicerone, "Chemically sensitive bioimaging with coherent Raman scattering," Nature Photonics 9, 295-305 (2015).

[8]. T. W. Kee, and M. T. Cicerone, "Simple approach to one-laser, broadband coherent anti-Stokes Raman scattering microscopy," Opt Lett 29, 2701-2703 (2004).

[9]. H. Kano, and H.-o. Hamaguchi, "Ultrabroadband (>2500cm^{-1}) multiplex coherent anti-Stokes Raman scattering microspectroscopy using a supercontinuum generated from a photonic crystal fiber," Applied Physics Letters 86, 121113 (2005).

[10]. T. Hellerer, A. M. K. Enejder, and A. Zumbusch, "Spectral focusing: High spectral resolution spectroscopy with broad-bandwidth laser pulses," Applied Physics Letters 85, 25-27 (2004).

[11]. I. Rocha-Mendoza, W. Langbein, and P. Borri, "Coherent anti-Stokes Raman microspectroscopy using spectral focusing with glass dispersion," Applied Physics Letters 93, 201103 (2008).

[12]. A. F. Pegoraro, A. Ridsdale, D. J. Moffatt, Y. Jia, J. P. Pezacki, and A. Stolow, "Optimally chirped multimodal CARS microscopy based on a single Ti:sapphire oscillator," Opt. Express 17, 2984-2996 (2009).

[13]. E. R. Andresen, P. Berto, and H. Rigneault, "Stimulated Raman scattering microscopy by spectral focusing and fiber-generated soliton as Stokes pulse," Opt Lett 36, 2387-2389 (2011).

[14]. H. T. Beier, G. D. Noojin, and B. A. Rockwell, "Stimulated Raman scattering using a single femtosecond oscillator with flexibility for imaging and spectral applications," Opt. Express 19, 18885-18892 (2011).

[15]. D. Fu, G. Holtom, C. Freudiger, X. Zhang, and X. S. Xie, "Hyperspectral Imaging with Stimulated Raman Scattering by Chirped Femtosecond Lasers," The Journal of Physical Chemistry B 117, 4634-4640 (2013).

[16]. J. G. Porquez, R. A. Cole, J. T. Tabarangao, and A. D. Slepkov, "Brighter CARS hypermicroscopy via "spectral surfing"of a Stokes supercontinuum," Opt Lett 42, 2255-2258 (2017).

[17]. F. Lu, and W. H. Knox, "Generation of a broadband continuum with high spectral coherence in tapered single-mode optical fibers," Opt. Express 12, 347-353 (2004).

[18] P. Abdolghader, A. F. Pegoraro, N. Y. Joly, A. Ridsdale, R. Lausten, F. Légaré, and A. Stolow, "All normal dispersion nonlinear fibre supercontinuum source characterization and application in hyperspectral stimulated Raman scattering microscopy," Opt. Express 28, 35997-36008 (2020).

[19] B. Figueroa, W. Fu, T. Nguyen, K. Shin, B. Manifold, F. Wise, and D. Fu, "Broadband hyperspectral stimulated Raman scattering microscopy with a parabolic fibre fibre fibre amplifier source," Biomed. Opt. Express 9(12), 6116–6131 (2018).

[20] D. Zhang, P. Wang, M. N. Slipchenko, D. Ben-Amotz, A. M. Weiner, and J. X. Cheng, "Quantitative vibrational imaging by hyperspectral stimulated Raman scattering microscopy and multivariate curve resolution analysis," Anal. Chem. 85(1), 98–106 (2013).
[21] Zhang X, de Juan A, Tauler R. "Multivariate Curve Resolution Applied to Hyperspectral Imaging Analysis of Chocolate Samples". Appl Spectrosc. 69(8),993-1003 (2015).

[22] Y. Ozeki, W. Umemura, Y. Otsuka, S. Satoh, H. Hashimoto, K. Sumimura, N. Nishizawa, K. Fukui, and K. Itoh, "High-speed molecular spectral imaging of tissue with stimulated Raman scattering," Nat. Photonics 6(12), 845–850 (2012).

[23] A. Alfonso-García, S. G. Pfisterer, H. Riezman, E. Ikonen, and E. O. Potma, "D38-cholesterol as a Raman active probe for imaging intracellular cholesterol storage," J. Biomed. Opt. 21(6), 061003 (2015).

[24] D. Fu and X. S. Xie, "Reliable cell segmentation based on spectral phasor analysis of hyperspectral stimulated Raman scattering imaging data," Anal. Chem. 86(9), 4115–4119 (2014).

[25] M. Wei, L. Shi, Y. Shen, Z. Zhao, A. Guzman, L. J. Kaufman, L. Wei, and W. Min, "Volumetric chemical imaging by clearing-enhanced stimulated Raman scattering microscopy," Proc. Natl. Acad. Sci. U.S.A. 116(14), 6608–6617 (2019).

[26] X. Zhang, M. B. J. Roeffaers, S. Basu, J. R. Daniele, D. Fu, C. W. Freudiger, G. R. Holtom, and X. S. Xie, "Label-free live-cell imaging of nucleic acids using stimulated Raman scattering microscopy," ChemPhysChem 13(4), 1054–1059 (2012).

[27] G. Gilboa, Y. Y. Zeevi, and N. Sochen, "Signal and image enhancement by a generalized forward-and-backward
adaptive diffusion process," in 2000 10th European Signal Processing Conference (2000), pp. 1–4.

[28] P. Perona and J. Malik, "Scale-space and edge detection using anisotropic diffusion," IEEE Trans. Pattern Anal. Mach. Intell. 12(7), 629–639 (1990).

[29] A. Buades, B. Coll, and J. Morel, "A non-local algorithm for image denoising," in 2005 IEEE Computer Society Conference on Computer Vision and Pattern Recognition (CVPR'05) (2005), 2, pp. 60–65 vol. 2.

[30] Elias Nehme, Lucien E. Weiss, Tomer Michaeli, and Yoav Shechtman, "Deep-STORM: super-resolution single-molecule microscopy by deep learning," Optica 5, 458-464 (2018).

[31] Yair Rivenson, Zoltán Göröcs, Harun Günaydin, Yibo Zhang, Hongda Wang, and Aydogan Ozcan, "Deep learning microscopy," Optica 4, 1437-1443 (2017).

[32] Sunil Kumar Gaire, Yang Zhang, Hongyu Li, Ray Yu, Hao F. Zhang, and Leslie Ying, "Accelerating multicolor spectroscopic single-molecule localization microscopy using deep learning," Biomed. Opt. Express 11, 2705-2721 (2020).

[33] Henry Pinkard, Zachary Phillips, Arman Babakhani, Daniel A. Fletcher, and Laura Waller, "Deep learning for single-shot autofocus microscopy," Optica 6, 794-797 (2019).

[34] Zhenxiang Luo, Abdulkadir Yurt, Richard Stahl, Andy Lambrechts, Veerle Reumers, Dries Braeken, and Liesbet Lagae, "Pixel super-resolution for lens-free holographic microscopy using deep learning neural networks," Opt. Express 27, 13531-13595 (2019).

[35] S. Weng, X. Xu, J. Li, and S. T. C. Wong, "Combining deep learning and coherent anti-Stokes Raman scattering imaging for automated differential diagnosis of lung cancer," J. Biomed. Opt. 22(10), 1–10 (2017).

[36] Alba Alfonso-García, Jerry Paugh, Marjan Farid, Sumit Garg, James Jester, and Eric Potma, "A machine learning framework to analyze hyperspectral stimulated Raman scattering microscopy images of expressed human meibum". *J. Raman Spectrosc.*, 48, 803– 812 (2017).

[37] Petru Manescu, Young Jong Lee, Charles Camp, Marcus Cicerone, Mary Brady, Peter Bajcsy, "Accurate and interpretable classification of microspectroscopy pixels using artificial neural networks", Med. Image. Analysis. 37, 37-45(2017).

[38] Olaf Ronneberger, Philipp Fischer, Thomas Brox, " U-net: Convolutional Networks for Biomedical Image Segmentation", arXiv:1505.04597

[39] Bryce Manifold, Elena Thomas, Andrew T. Francis, Andrew H. Hill, and Dan Fu, "Denoising of stimulated Raman scattering microscopy images via deep learning," Biomed. Opt. Express 10, 3860-3874 (2019).

[40] Haonan Lin, Hyeon Jeong Lee, Nathan Tague, Jean-Baptiste Lugagne, Cheng Zong, Fengyuan Deng, Wilson Wong, Mary J. Dunlop, Ji-Xin Cheng, "Fingerprint Spectroscopic SRS Imaging of Single Living Cells and Whole Brain by Ultrafast Tuning and Spatial-Spectral Learning", arXiv:2003.02224, (2020).

[41] Jing Zhang, Jian Zhao, Haonan Lin, Yuying Tan, and Ji-Xin Cheng, "High-Speed Chemical Imaging by Dense-Net Learning of Femtosecond Stimulated Raman Scattering", J. Phys. Chem. Lett, 11, 8573–8578, (2020)

[42] T. A. Pologrutc, B. L. Sabatini, and K. Svoboda, Biomed. Eng. Online 2,13 (2003).

[43] Diederik P. Kirgma, Jimmy Ba, "ADAM: A METHOD FOR STOCHASTIC OPTIMIZATION", https://arxiv.org/abs/1412.6980

Chapter 5

5. Conclusion

Coherent Raman microscopy (CRM), specifically SRS microscopy, is a cutting-edge nonlinear optical imaging technique which opened doors in many fields where Raman hyperspectral imaging is necessary. Molecular vibrations are commonly used to achieve chemical contrast with this technique. In SRS microscopy, the use of two ultrashort (femtosecond) laser sources with low noise characteristics combined with a rapid broadband tunability is a benefit. When other nonlinear effects such as two-photon absorption, cross-phase modulation, and thermal lensing are suppressed, sample identification based on Raman spectroscopy is achievable in SRS microscopy. The use of a broadband laser source allows for the acquisition of a broad Raman spectrum per pixel, aiding sample classification-based pattern identification of the recorded Raman spectrum. We note that SNR values of recorded images can decline considerably in some imaging conditions, such as real-time or deep tissue imaging.

In this thesis, we began with a consideration of the laser source. Nonlinear fibres operating in the anomalous dispersion regime produce noisy output, making them unsuitable for SRS microscopy. We introduced a new fibre light source with noise performance adequate for SRS microscopy: the All Normal Dispersion (ANDi) regime was used in the design this new fibre source. The particular dispersion profile of this fibre source is the essential feature. Soliton dynamics can induce large intensity variations in ordinary supercontinuum generation (SCG) fibre sources because the generated supercontinuum is strongly dependent on input laser characteristics. In the ANDi regime, however, SCG solely arises from self-phase modulation. We measured the noise in an ANDi fibre to characterize its continuum generation in two frequency regimes: (i) the inverse of pixel dwell time (10-40) KHz, and (ii) the Stokes modulation frequency ($\sim$ 1MHz). The ANDi fibre's continuum was used to scan two samples: neat dimethyl sulfoxide (DMSO) liquid and acetaminophen powder. Using the ANDi supercontinuum provided a $\sim$ 800 cm^{-1} Raman scan range without any need for tuning the laser.

In the second part of this thesis, we considered image processing methods for extracting quantitative information from hyperspectral SRS images. Recently, Deep Learning has emerged as a powerful data analysis technique, demonstrating remarkable performance in imaging applications such as cancer detection and super-resolution microscopy. Both image denoising and sample classification were successfully achieved using supervised deep learning algorithms in CRM. However, the majority of these studies necessitated the use of a large datasets and ground-truth images. We argued here that a single hyperspectral SRS image with a low SNR would be sufficient to train a customized model in an unsupervised approach.

A typical unsupervised machine learning neural network topography is an autoencoder. Autoencoders are based on the concept of an information bottleneck which forces a network to compress a signal into a low-dimensional space. The capacity to project high-dimensional data (hyperspectral images) into a low-dimensional space based on correlations and similarities identified within an unsupervised, unlabelled training set is a significant capacity demonstrated here: it is a feature (or representation) learning example. Our new approach, UHRED (Unsupervised Hyperspectral Resolution Enhancement and Denoising), is label-free and requires no "ground truth data" having a high SNR. Furthermore, we demonstrated that UHRED could automatically segment and identify various features within a dataset without the need for human intervention.

We created synthetic datasets and training our customized convolutional autoencoder model with various SNR input signal conditions. In order to achieve optimal training parameters, model parameters such as the number of layers, optimizer step size, and the number of iterations were varied. We picked two samples and experimentally recorded the hyperspectral nonlinear optical spectral signals (termed here the hyperspectral index) at each image pixel. Importantly, the definition of the hyperspectral index is quite broad and can represent any linear or nonlinear optical signal channels, such as SRS, Pump-Probe, thermal lensing, cross-phase modulation microscopy, or any other signal types which rely on a laser parameter as input. We demomnstrated that our proposed model automatically denoised and segmented images. UHRED is a label-free, unsupervised deep learning approach which: (i) permits reduction in either the optical power requirements or integration time needed to achieve a target SNR; (ii) can work with training data collected only under low-SNR conditions (requiring no high-SNR data); (iii) offers unsupervised identification and segmentation of spectrally distinct materials based on a full (multimodal) hyperspectral data set. The UHRED approach improves contrast and Raman spectral resolution automatically, as well as providing complete segmentation for chemical (mineral species) maps.

We could imagine improving the performance of our model by adding additional imaging channels, such as a second harmonic generation or coherent anti-Stokes Raman scattering channels. Every model input is treated as a feature vector in Deep Learning, thus improving the model's accuracy. It may also be preferable to use supervised learning algorithms in some circumstances where high SNR images are available. Real-time imaging is another area that we believe has a huge potential. The use of Deep Learning models in conjunction with the CRM approach could result in a new imaging capacity in clinical applications.

www.ingramcontent.com/pod-product-compliance
Lightning Source LLC
LaVergne TN
LVHW041701190726
843493LV00007B/1906

9 783384 271488